New Delivery Systems for Controlled Drug Release from Naturally Occurring Materials

About the Cover

The book cover demonstrates the effect of pH and temperature on sol–gel phase transition of block copolymer solutions. *See* Chapter 7, Figure 7. (Drawing courtesy of Doo-Sung Lee, Woo Sun Shim, and Dai Phu Huynh.)

ACS SYMPOSIUM SERIES **992**

New Delivery Systems for Controlled Drug Release from Naturally Occurring Materials

Nicholas Parris, EDITOR
Agricultural Research Service, U.S. Department of Agriculture

LinShu Liu, EDITOR
Agricultural Research Service, U.S. Department of Agriculture

Cunxian Song, EDITOR
Chinese Academy of Medical Sciences

V. Prasad Shastri, EDITOR
Vanderbilt University

Sponsored by the
ACS Division of Macromolecular Chemistry

American Chemical Society, Washington, DC

chem
scplae

ISBN: 978–0–8412–7424–2

The paper used in this publication meets the minimum requirements of American National Standard for Information Sciences—Permanence of Paper for Printed Library Materials, ANSI Z39.48–1984.

PRINTED IN THE UNITED STATES OF AMERICA

1/6/09 KA

Foreword

The ACS Symposium Series was first published in 1974 to provide a mechanism for publishing symposia quickly in book form. The purpose of the series is to publish timely, comprehensive books developed from ACS sponsored symposia based on current scientific research. Occasionally, books are developed from symposia sponsored by other organizations when the topic is of keen interest to the chemistry audience.

Before agreeing to publish a book, the proposed table of contents is reviewed for appropriate and comprehensive coverage and for interest to the audience. Some papers may be excluded to better focus the book; others may be added to provide comprehensiveness. When appropriate, overview or introductory chapters are added. Drafts of chapters are peer-reviewed prior to final acceptance or rejection, and manuscripts are prepared in camera-ready format.

As a rule, only original research papers and original review papers are included in the volumes. Verbatim reproductions of previously published papers are not accepted.

ACS Books Department

Contents

Materials for Drug Delivery Systems

Drug Delivery Systems

Agricultural and Food Applications and Analytical Methodologies

Indexes

Preface

Controlled drug delivery systems are prepared from synthetic or natural materials and combined with a drug or an active ingredient allowing its release in a predetermined fashion. The release can be constant, cyclic, or triggered by changes in physiological environment or an external event. The development of novel drug delivery systems that are capable of controlling the release of drugs to specific sites in the body has attracted considerable attention. Early biomaterials were selected primarily for their desirable physical properties and not for delivery of active ingredients in biological systems. In recent years, however, new systems have been prepared from naturally occurring materials that are biocompatible and resistant to enzymatic and chemical attack in the in vivo environment.

Today, microencapsulation and controlled-release technologies are finding broad application not only in the pharmaceutical industry but also for transplantation and regeneration medicine, for example, scaffold-based drug delivery for tissue engineering. The advantages of using controlled delivery systems in the aforementioned applications is primarily the maintenance of drug levels within defined ranges, in a localized manner while maintaining overall biocompatibility. Possible disadvantages, however, include the high cost of controlled-release systems, potential toxicity, or non-biocompatibility of the materials used.

This book is targeted toward scientists and engineers in industry, government, and academia who are interested in the development of novel drug delivery systems and regeneration technologies from natural materials. An understanding of physicochemical changes and how such changes affect the performance of the drug or active ingredient will allow researchers to develop formulations with optimized performance.

This volume was developed from a symposium presented at the 2005 International Chemical Congress of Pacific Basin Societies sponsored by the American Chemical Society Division of Macromolecular Chemistry in Honolulu, Hawaii, December 15–20, 2005. The editors attribute the success of the symposium to the selection of speakers who are recognized experts from both industry and academia. To those experts who were missed we offer our apologies. We thank the speakers for their excellent presentations and the timely preparation of their chapters. We also thank the reviewers for their incisive and constructive criticism.

Nicholas Parris
Eastern Regional Research Center
Agricultural Research Service
U.S. Department of Agriculture
600 East Mermaid Lane
Wyndmoor Lane, PA 19038
nicholas.parris@ars.usda.gov

LinShu Liu
Eastern Regional Research Center
Agricultural Research Service
U.S. Department of Agriculture
600 East Mermaid Lane
Wyndmoor Lane, PA 19038
linshu.liu@ars.usda.gov

Cunxian Song
Institute of Biomedical Engineering
Chinese Academy of Medical Sciences
Tianjin 300192
People's Republic of China
scxian@tom.com

V. Prasad Shastri
Department of Biomedical Engineering
Vanderbilt University
5824 Stevenson Center
Nashville, TN 37235
Prasad.shastri@vanderbilt.edu

New Delivery Systems
for Controlled Drug Release
from Naturally Occurring
Materials

Materials for Drug Delivery Systems

Chapter 1

Efficacy Assessment of Hyaluronan Vehicles for Sustained Release of pDNA

Weiliam Chen[1,2], Angela Kim[2], Mary Frame[1], Daniel Checkla[2], Yang-Hyun Yun[1], Paige Yellen[1], and Philip Dehazya[2]

[1]Department of Biomedical Engineering, State University of New York-Stony Brook, Stony Brook, NY 11794-2580
[2]Clear Solutions, Inc., Stony Brook, NY 11790-3350

Natural polymers have many advantages over their synthetic counterparts for biomedical uses; carbohydrate is a class of under-explored natural polymers for gene delivery. We have formulated microspheres and wafer matrices from Hyaluronan (HA) for delivery of plasmid DNA (pDNA). Sustained releases of pDNA over a period of at least 1 month were achieved in both formulations. Results from cell culture transfection studies indicated that the pDNA released from these systems were both intact and bioactive. Microsphere formulations for sustained delivery of pDNA encoding Vascular Endothelial Growth Factor (VEGF) were infiltrated into adult hamster cheek pouches, the microvascular responses differed from control. Wafer matrix formulations for sustained release of pDNA encoding Platelet Derived Growth Factor-BB (PDGF-BB) were implanted in pig full-thickness dermal wound models and the results indicated acceleration of wound repair.

Introduction

Natural carbohydrate, a class of generally recognized safe biomaterials with low toxicity potential, is under-explored as delivery vehicles in the gene therapy arena. Hyaluronan (HA) is a high molecular weight glycosaminoglycan with exceptional biocompatibility and is virtually non-immunogenic. The abundance of carboxyl groups on HA enables it to be crosslinked with adipic dihydrazide rendering it insoluble, while maintaining the biocompatibility, biodegradability, and non-immunogenicity characteristics of native HA (*1-4*). The general reaction scheme is depicted in Figure 1. The dihydrazide crosslinking chemistry is aqueous based and all reagents and reaction byproducts are non-toxic.

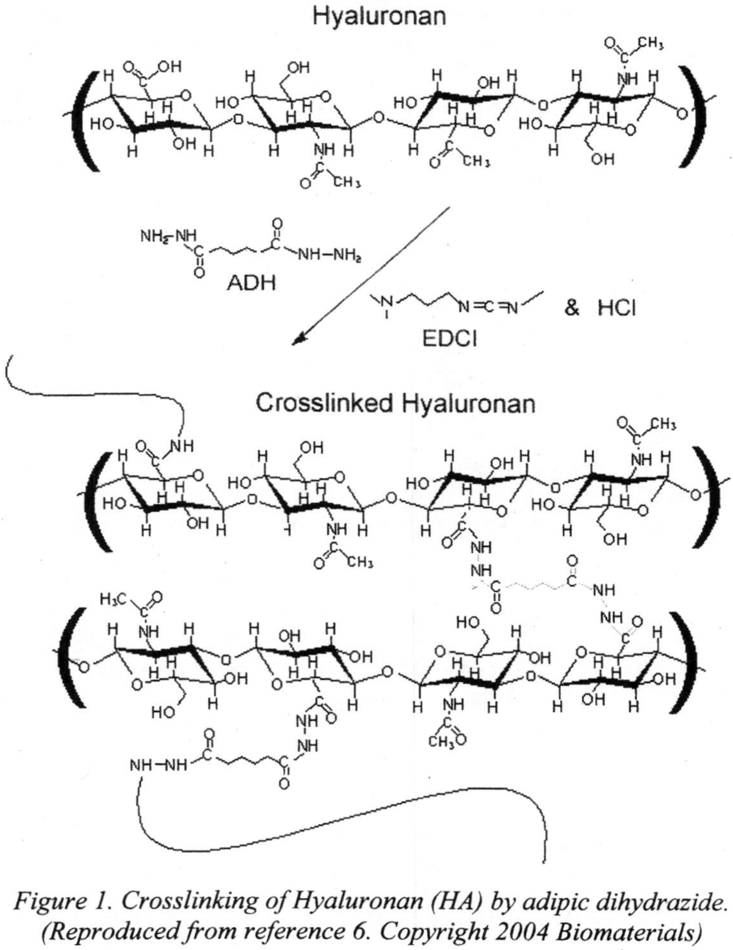

Figure 1. Crosslinking of Hyaluronan (HA) by adipic dihydrazide. (Reproduced from reference 6. Copyright 2004 Biomaterials)

Naked plasmid DNA (pDNA) as a vector for gene transfer faces many obstacles such as low transfection efficiency, limited duration of activity and susceptibility to degradation by nuclease (5). Coupling pDNA to HA is potentially a strategy to address these issues; moreover, it could also overcome the toxicity concerns associated with using viral vectors and many experimental polycations utilized for gene delivery (5).

In this chapter, we describe two HA based vehicles, microspheres and matrix, capable of sustained release of pDNA and the release kinetics can be modulated. Continued releases of pDNA greater than 1 month were achieved in both vehicles. pDNAs encoding β-galactosidase and PDGF-BB-Flag fusion gene were released intact from microspheres and matrices, respectively; cell transfection indicated that their bioactivities were intact. Clinically pertinent pDNAs encoding VEGF and PDGF-BB were incorporated into these vehicles and they were tested in animal models. Prolonged localized VEGF activity was demonstrated after HA microspheres containing pDNA encoding VEGF were injected into hamster cheek pouches. Acceleration of wound repair was detected in porcine full-thickness dermal wound model after HA matrix containing pDNA encoding PDGF-BB were implanted.

Materials and Methods

Formulation of Microspheres and Matrices

DNA-HA microspheres were formulated using a modified in-emulsion crosslinking method described previously (6). Briefly, a W/O emulsion was formed by homogenizing 20 mL of 0.5% hyaluronan (Kraeber GMBH & Co., Waldhofstr, Germany) solution containing 100 mg of dissolved adipic dihydrazide (Sigma-Aldrich, St. Louis, MO), and 80 mL of mineral oil (Sigma-Aldrich, St. Louis, MO) with 1 ml of Span 80 (ICI Chemicals, Wilmington, DE) dissolved, at 1,000 RPM. A pDNA solution (1 mL, with 1 or 5 mg of pDNA dissolved in it) was added dropwise; thereafter, an ethyl-3-[3-dimethyl amino] propyl carbodiimide (Sigma-Aldrich, St. Louis, MO) solution (120 mg dissolved in 2 mL of water) was instilled into the emulsion. This was followed by the addition of 0.3 mL of 0.1N HCl to initiate crosslinking reaction. The HA-DNA microspheres were recovered by adding 150 ml of isopropyl alcohol to the emulsion followed by centrifugation at 1,500 RPM and lyophilization (Virtis Freezemobile 12EL, Gardener, NY).

DNA-HA matrices were formulated using a method described previously (7). Briefly, DNA-HA matrices were prepared by first lyophilizing a solution of 20 mL 1% HA solution with 1 to 5 mg of pDNA dissolved in it. The lyophilized foam was then immersed in 100 mL of a reagent solution of adipic dihydrazide (100 mg) dissolved in 90% isopropyl alcohol (IPA)/10% water for 30 minutes.

This was followed by the addition of 120 mg of ethyl-3-[3-dimethyl amino] propyl carbodiimide. Upon complete dissolution, crosslinking reaction was initiated by acidifying the reagent solution to pH 5 with 1N HCl, while under gentle agitation. After subsequent incubation for 5 to 25 hours at ambient temperature, the reagent was replaced by 100 ml of 90% IPA/10% water. Residual reagents were removed by several additional washes followed by 100% IPA. The IPA was decanted and its residual removed by aspiration.

Release of pDNA from Microspheres and Matrices and Gel Electrophoresis

The pDNA released from DNA-HA microspheres were evaluated. DNA-HA microspheres were dispersed in 1 ml of hyaluronidase solution (Sigma-Aldrich, St. Louis, MO, 10 units/mL in pH 7.4 phosphate buffered saline) (8). At pre-determined sampling time-point, the DNA-HA microsphere suspensions were centrifuged at 3,000 RPM and 0.9 ml of the supernatants was recovered. An equal volume of hyaluronidase solution was added to re-suspend the microspheres. Likewise, the DNA release kinetics of crosslinked DNA-HA matrices were evaluated. Briefly, matrices were hydrated in 1.5 mL of hyaluronidase solution and placed under constant agitation at 37°C. At each sampling time-point, the containers were centrifuged at 4,000 RPM and 1 mL of sample was collected from each tube, and the DNA-HA matrices were re-suspended in an equal volume of hyaluronidase solution. A Picogreen fluorometric assay) was adopted to determine the pDNA concentration of the supernatants collected during the course of the release studies. Gel electrophoreses were performed on pDNA samples using agarose gels (6, 7).

Assessment of pDNA Activity Collected from Release Studies

DNA-HA microspheres were formulated using pDNA encoding β-galactosidase. The bioactivity of pDNA released were evaluated with CHO cells (ATCC, Rockville, MD) using established LIPOFECTAMINE™ (Gibco Invitrogen, Carlsbad, CA) mediated protocol (6). The extents of transfection were evaluated by cytochemical analyses (X-Gal™ staining, Gibco Invitrogen, Carlsbad, CA) and the numbers of transfected cells were manually counted.

DNA-HA matrices were formulated with pDNA encoding PDGF-BB or PDGF-BB-Flag (Flag™ is a commercially available 24-nucleotide sequence, encoding an octapeptide epitope that was attached to the terminal nucleotide encoding PDGF-BB to facilitate verification of its exogenous origin) (Flag™ Expression kit, Sigma-Aldrich) incorporated and subjected to digestion in media containing hyaluronidase. pDNA encoding PDGF-BB-Flag released from DNA-HA matrices were used to transfect COS-1 cells and they were stained with anti-Flag antibodies (7).

The efficacy of the PDGF-BB pDNA released in stimulating normal neonatal human dermal fibroblast (NNHDF) was assessed. Briefly, COS-1 cells were transfected with randomly selected PDGF-BB pDNA releasates (days 3, 10, 17 and 24) with established protocols (7). These cells were sustained on FGM-2 BulletKits® medium with reduced serum concentration (0.5%) and without the growth supplements (i.e., a starving medium). Three days after transfection, the media were collected and diluted (1:2) with some fresh modified FGM-2 BulletKits® medium. The blend was used to induce proliferation of NNHDF (seeded at 5×10^4 cells per cm^2), and the cell densities were monitored for 5 consecutive days. The cell densities on individual plates were determined by the level of fluorescent signals produced with a Hoescht 33342 (Molecular Probes, Eugene, OR) mediated DNA assay (7). β-galactosidase pDNA transfected cells were used as controls

In Vivo Validation of Gene Expression and Efficacies

DNA-HA microspheres (containing pDNA encoding β-galactosidase) suspension (0.5 mg of microspheres dispersed in 50 μL of saline) was injected into rat hind limb muscles. Contra-lateral limbs were injected with either negative or positive (conventional poly-lactide-co-glycolide microspheres with comparable pDNA loading) controls (10). Three weeks post-injections, rats were euthanized and the muscles were promptly retrieved and frozen in liquid nitrogen. The frozen specimens were reduced to powder and transferred into RNase free centrifuge tubes containing Trizol® (Gibco Invitrogen, Carlsbad, CA), and homogenized (PowerGen 700, Fisher Scientific, Pittsburgh, PA). The RNA sample was then extracted, precipitated, washed and purified utilizing commercially available reagent kits following manufacturers' established protocols. The extracted RNA was analyzed by reverse transcriptase polymerase chain reaction (RT-PCR) utilizing a RT-PCR assay kit (Qiagen Inc., Valencia, CA) and the amplicon was sequenced to ensure it was originated from the β-galactosidase.

Using a hamster cheek pouch model, DNA-HA microspheres (containing pDNA encoding VEGF) suspension were infiltrated into the space behind the cheek pouch, using a published technique (11). Briefly, the cheek pouch was exteriorized for intravital microscopy and continuously superfused with 7.4 pH physiological salt solution (in mmol/l: NaCl 128, KCl 4.7, $CaCl_2$ $2H_2O$ 2.0, $MgSO_4$ $7H_2O$ 1.2, $NaHCO_3$ 17.8, EDTA 0.02, NaH_2PO_4 H_2O 1.2, glucose 5.0, pyruvate 2.0). Nine days after microsphere infiltration, arteriolar vasoactive responses were tested in the hamster cheek pouch preparation (12) and compared to laboratory controls or to juvenile hamsters (not treated).

A porcine full-thickness dermal wound model was utilized to assess the efficacy of the DNA-HA matrices (containing pDNA encoding PDGF). Under

8

general anesthesia, the animals were shaved and aseptically prepped; 8-mm full-thickness wounds were created with biopsy punches. Each wound bed received a DNA-HA matrix followed by plastering of Tegaderm™ over the wound bed. At day 4 or 11, the animals were euthanized and their relevant wound sites were excised, paraffin embedded and processed for histology analyses after H&E staining. All specimens were evaluated with NIH Image J software for their extent of granulation tissue formation using NIH Image J.

Results and Discussion

DNA Release Kinetics from Microspheres and Matrices

The appearance of DNA-HA microspheres was shown in Figure 2. The DNA release kinetics of two typical microsphere preparations (pDNA loading 1 and 5%, respectively) were depicted in Figure 3. Approximately 60% of the microspheres' pDNA contents were released at the conclusion of the study (63 days). The presence of hyaluronidase was needed to digest the DNA-HA microspheres to release the pDNA. Figure 4 depicted the gel electrophoresis results of some pDNA releasate samples and they were intact (lanes 3 to 6). The fluorescing smears along the migration paths suggested the presence of pDNA conjugated to crosslinked HA fragments of various sizes.

The appearance of DNA-HA matrices was shown in Figure 5. The pDNA release kinetics of two typical matrix formulations (pDNA loading 0.5% and 2.5%, respectively) were depicted in Figure 6. Approximately 50% of the matrices' DNA contents were released at the conclusion of the study (37 days).

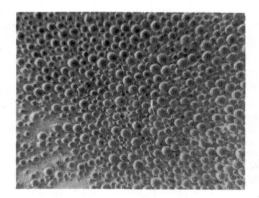

Figure 2. DNA-HA Microspheres. (Reproduced from reference 13. Copyright 2005 Polymeric Gene Delivery Vehicles: Principles and Applications, CRC)

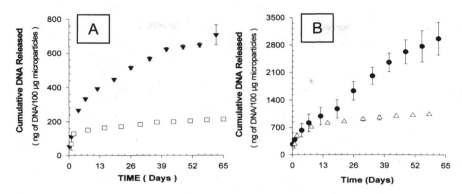

Figure 3. Sustained release of pDNA from DNA-HA microspheres formulations. *(A) 1% pDNA loading, ▼ in hyaluronaidase, and □ in PBS. (B) 5% pDNA loading, ● in hyaluronidase, and △ in PBS. (Reproduced from reference 6. Copyright 2004 Biomaterials)*

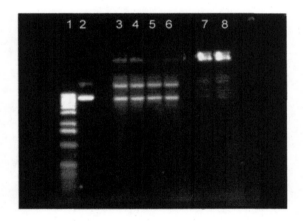

Figure 4. Gel electrophoresis of pDNA released from DNA-HA microspheres formulations. *Gel electrophoresis: (1) DNA ladder, (2) pDNA for formulating DNA-HA microspheres, (3,4,5,6) pDNA releasates collected, (7) DNA-HA microspheres containing 1% pDNA, and (8) DNA-HA microspheres containing 5% pDNA. (Reproduced from reference 6. Copyright 2004 Biomaterials)*

*Figure 5. DNA-HA matrices. (Reproduced from reference 7.
Copyright 2003 Journal of Controlled Release)*

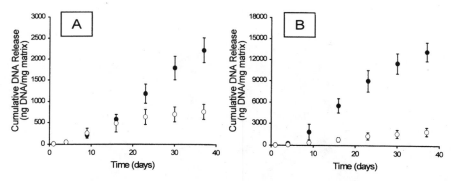

*Figure 6. Sustained release of pDNA from DNA-HA matrices. (A) 0.5% pDNA
loading. (B) 2.5% pDNA loading. ● in hyaluronidase, and ○ in PBS.
(Reproduced from reference 7. Copyright 2003 Journal of Controlled Release)*

Figure 7 depicted the gel electrophoresis results of some pDNA releasate
samples and they were intact. Similar to the results depicted above (Figure 4),
there were fluorescing smears along the migration paths in addition to intense
fluorescences inside the wells, this also suggested the presence of pDNA
conjugated to crosslinked HA fragments of various sizes, with the largest one
incapable of migrating into the agarose gel.

Day 9   Day 16   Day 23   Day 30   Day 37

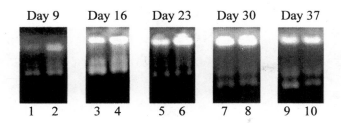

1   2   3   4   5   6   7   8   9   10

Figure 7. Electrophoretic motility analyses of pDNA released from DNA-HA matrices. (1, 3, 5, 7, 9): from matrices containing 2.5% pDNA, crosslinked for 5 hours. (2, 4, 6, 8, 10): pDNA from matrices containing 0.5% DNA crosslinked for 25 hours. (Reproduced from reference 14. Copyright 2005 Polymeric Gene Delivery Vehicles: Principles and Applications, CRC)

Validation of Bioactivities by In Vitro Transfection in Cell Culture Models

Figure 8 summarized the relative levels of transfection in CHO cells using releasates collected from the DNA-HA microsphere formulation containing 1% of pDNA (encoding β-galactosidase) loading. Overall, the relative levels of transfection mimics the DNA release profile.

pDNA (encoding PDGF-BB-Flag fusion gene) releasates collected from DNA-HA matrices were used for transfecting COS-1 cells and subjected to immunofluorescent staining utilizing anti-Flag epitope antibody. Sample results utilizing pDNA releasates collected on Days 7, 17 and 41 for transfection were shown in Figure 9. The apparent magnitudes of transfection (as indicated by the

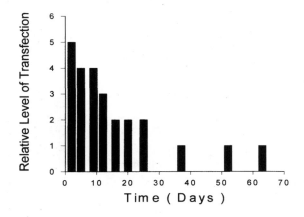

Figure 8. Relative levels of transfection using pDNA released from DNA-HA microspheres. (Reproduced from reference 6. Copyright 2004 Biomaterials)

A (Day 7)   B (Day 17)   C (Day 41)

D (Day 7)   E (Day 17)   F (Day 41)

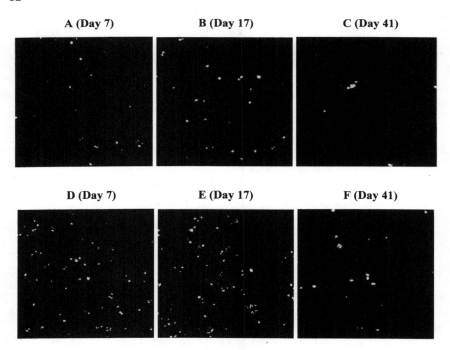

Figure 9. Immunofluorescent staining of cells transfected with pDNA, encoding PDGF-BB-Flag, released from DNA-HA matrices. (A, B, C) from matrix with 0.5% pDNA loading. (D, E, F) from matrix with 2.0% pDNA loading. (Reproduced from reference 7. Copyright 2003 Journal of Controlled Release)

density of fluorescing cells) qualitatively correlated with the original pDNA contents of the matrices and thus, the pDNA concentrations of the releasates.

The results of the cell proliferation study was depicted in Figure 10. Evidently NNHDF, when exposed to the PDGF-BB conditioned starving media, exhibited a much greater degree of proliferation when compared to those exposed to starving media conditioned with β-galactosidase. This remained consistent with all releasates tested (days 3, 10, 17 and 24). By day 3, cells conditioned with the media containing PDGF-BB were approximately twice as confluent as the cells conditioned with β-galactosidase; and evidently, the proliferation had reached plateau due to exhaustion of nutrients. Cells appeared to proliferate modestly in β-galactosidase conditioned media and the negative control.

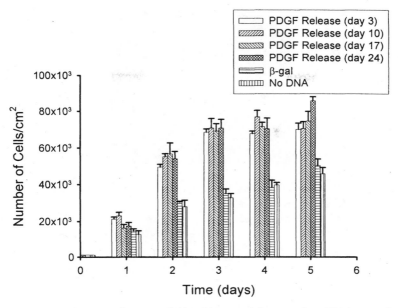

Figure 10. Induction of neonatal fibroblast proliferation by pDNA, encoding PDGF-BB, released from DNA-HA matrices. (Reproduced from reference 7. Copyright 2003 Journal of Controlled Release)

In Vivo Validation of Gene Expression and Efficacies

The RT-PCR results of rat hind limb muscles three weeks after DNA-HA microsphere injection were depicted in Figure 11. Lanes 3 and 4 were samples from two animals injected with DNA-HA microspheres; whereas lane 5 (a faint signal) was produced with the sample injected with PLGA microspheres containing pDNA encoding β-galactosidase). No signal was detected on the sample injected with saline (lane 2). Lane 6 was the β-galactosidase DNA template as a positive control.

Nine days after DNA-HA microsphere infiltration, hamster cheek pouch arteriolar vasoactive responses were tested (*12*). The results were summarized in Table I. Importantly, these microspheres did not appear to have any adverse effects on the animals. Baseline diameters were not different for the three groups: the average baseline diameter was 9.18 ± 2.62 µm. Diameter change was examined as a fractional change from baseline in response to micropipette delivered test agents (*12*). Non-receptor mediated constriction (20 mM KCl) and receptor mediated constriction (10^{-8} M endothelin) were not different in the three groups. However, both non-receptor mediated dilation (10^{-4} M nitroprusside)

14

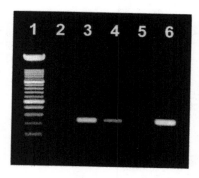

Figure 11. RT-PCR of rat hind limbs injected with DNA-HA microspheres, saline, or PLGA microspheres. (1) DNA ladder, (2) saline, (3, 4) DNA-HA microspheres, (5) PLGA microspheres, and (6) positive control. (Reproduced from reference 6. Copyright 2004 Biomaterials)

Table I. Vasoactive Responses of Hamster Cheek Pouches (mean±SD).

	Control	*DNA-HA microspheres*	*Juvenile*
	(n=11)	(n=5)	(n=13)
Basal diameter (μm)	8.63±2.66	9.47±2.27	9.11±2.71
KCl	-0.25±0.20	-0.16±0.06	-0.22±0.05
Endothelin	-0.45±0.10	-0.37±0.13	-0.37±0.07
SNP local	1.00±0.10	0.35±0.19	0.48±0.07
ADO local	0.92±0.10	0.40±0.11	0.54±0.05
SNP remote	0.64±0.20	0.27±0.07	0.36±0.04
LM609	0.62±0.10	0.22±0.14	0.19±0.02
Preconditioning	0.88±0.10	0.27±0.05	0.36±0.05

Vasoactive responses, reported as a fractional change from baseline diameter. Multiply by 100 to get the %-change in diameter.

and receptor mediated dilation (10^{-4} M adenosine) were significantly diminished in the juvenile, and DNA-HA microspheres treated groups as compared to adult hamsters. We also examined vascular communication responses in which the micropipette applied test agent was applied remotely 1000 μm downstream from the upstream observation arteriole (*12, 15*). These types of responses tested the ability of the vasculature to communicate and coordinate vascular resistance changes. Remote vasodilation to nitroprusside (10^{-4} M) was suppressed in both the DNA-HA microspheres treated and juvenile animals as compared to adult controls. Remote vasodilation to the alpha V beta 3 integrin ligating agent, LM609 (10 μg/mL) consists of both a flow mediated dilation and dilation requiring gap junctional activity (*15, 16*). The remote dilation to LM609 was decreased by half in the DNA-HA microspheres and juvenile groups compared to adult controls. Lastly we tested microvascular preconditioning, as before using 10^{-4} M nitroprusside to precondition the arterioles, and testing for dilation to the amino acid, L-arginine (10^{-4} M) (*17, 18*). Preconditioning is normally seen as a significant dilation to micropipette applied L-arginine in adult hamsters. This response was significantly decreased in both the DNA-HA microspheres treated and juvenile groups as compared to adult controls. Thus, while the constrictor responses tested were not different for the chronic VEGF DNA treated animals as compared to adults controls, several types of dilatory responses were diminished by this treatment. The responses with chronic VEGF treatment were similar to those found in juvenile hamsters.

The histology specimens of full thickness dermal wounds (pig model) were evaluated for percent granulation tissue formation. The results were summarized in Table II. Overall, the wounds implanted with DNA-HA matrices containing pDNA encoding PDGF-BB showed greater extent of granulation tissue formation.

Table II. Percent Granulation Tissue Formation in Full-Thickness Wound Implanted with DNA-HA Matrices. The Values Represent Average±SD. The Number in Parentheses are p-Values.

PDGF-BB	*β-Galactosidase*	*No. DNA*
72.0±13.5	46.1±7.5 (0.003)	55.8±3.4 (0.014)

References

1. Vercruysse, K. P.; Marecak, D. M.; Marecek, J. F.; Prestwich, G. D. *Bioconj. Chem.* **1997**, *8*, 686.
2. Pouyani, T.; Prestwich, G. D. *Bioconj. Chem.* **1994**, *5*, 339.
3. Pouyani, T.; Prestwich, G. D. *Bioconj. Chem.* **1994**, *5*, 370.
4. Pouyani, T.; Harbison, G. S.; Prestwich, G. D. *J. Am. Chem. Soc.* **1994**, *116*, 7515.
5. Luo, D. D.; Saltzman, W. M. *Nat. Biotechnol.* **2000**, *18*, 33.
6. Yun, Y. H.; Goetz, D.; Yellen, P.; Chen, W. *Biomaterials.* **2004**, *25*, 147.
7. Kim, A.; Checkla, D.; Dehazya, P.; Chen, W. *J. Controlled Rel.* **2003**, *90*, 81.
8. Delpech, B.; Bertrand, P.; Chauzy, C. *J. Immunol. Methods* **1987**, *104*, 223.
9. Labarca, C.; Paigen, K. *Anal. Biochem.* **1980**, *102*, 344.
10. Wang, D.; Robinson, D. R.; Kwon, G. S.; Samuel, J. *J. Control Release.* **1999**, *57*, 9.
11. Frame, M. D.; Miano, J. M.; Yang, J.; Rivers, R. J. *Microcirculation* **2001**, *8*, 403.

12. Rivers, R. J.; Frame, M. D. S. *J. Vasc. Res.* **1999**, *36*, 193.
13. *Polymeric Gene Delivery Vehicles: Principles and Applications;* Amiji, M. M., Eds., CRC Press, Baton Rouge, 2005; p 474.
14. *Polymeric Gene Delivery Vehicles: Principles and Applications;* Amiji, M. M., Eds., CRC Press: Baton Rouge, 2005; p 589.
15. Frame, M. D. *Am. J. Physiol. Heart Circ.* **2000**, *278*, H1186.
16. Frame, M. D. S. *Am. J. Physiol. Heart Circ.* **1999**, *276*, H1012.
17. Mabanta, L.; Valane, P.; Borne, J.; Frame, M. D. *Am. J. Physiol. Heart Circ.* **2006**, *290*, H264.
18. Frame, M. D. S.; Sarelius, L. H. *Circ. Res.* **1995**, *77*, 695.

Chapter 2

A New Soy-Based Hydrogels: Development, Viscoelastic Properties, and Application for Controlled Drug Release

Zengshe Liu and Sevim, Z. Erhan

Food and Industrial Oil Research, National Center for Agricultural Utilization Research, Agricultural Research Service, U.S. Department of Agriculture, 1815 North University Street, Peoria, IL 61604

Polymeric drug delivery systems have attracted increasing attention during the last two decades. Amphiphilic block copolymers have been widely studied due to their potential application in drug delivery systems as they are capable of forming aggregates in aqueous solutions. These aggregates comprise of a hydrophobic core and hydrophilic shell. They are good vehicles for delivering hydrophobic drugs, since the drugs are protected from possible degradation by enzymes (*1-10*).

Temperature-gelling PEO-PPO-PEO triblock copolymers of poly(ethylene oxide) (PEO) and poly(propylene oxide) (PPO), known as a Pluronic surfactant, are used as solubilization agents in pharmaceutical applications (*11*). This macromolecular surfactant has also been extensively studied as a potential drug delivery vehicle due to its excellent biocompatibility (*5*). It is one of the very few synthetic polymeric materials approved by the U.S. Food and Drug Administration for use as a food additive and pharmaceutical ingredient.

However, the critical micelle concentration (CMC) of Pluronic block copolymers is typically very high because of the low hydrophobicity of PPO blocks. This limits the application of Pluronic micelles because they are not stable and are easily destroyed by dilution when injected into the human body. It is for this reason that chemical modifications of Pluronic block polymers may be necessary. For example, Pluronic copolymers were hydrophobically modified with polycaprolactone (PCL). The modified block copolymer of PCL-Pluronic-PCL possessed a much lower CMC due to the hydrophobicity of PCL segments (*12, 13*). Other researchers have grafted Pluronic with poly(acrylic acid) (PAA) and poly(lactic acid)(PLA). The gelation concentration of modified copolymers is much lower than that of Pluronic (*14-17*). Such graft copolymer structures combine bioadhesive and hydrophobic properties in a single molecule, retain thermoreversible gelation behavior over a wide pH range, and do not permit physical separation. Therefore, they can be applied in vaginal drug delivery (*18-20*). A striking feature of Pluronic-PAA graft-copolymer is the ability to form the thermogel at low polymer concentrations, when neither parent Pluronic nor 1:1 physical blend of Pluronic and poly(acrylic acid) show any signs of gelation (*19*).

Hydrogels as drug carriers are able to provide a set of advantages. This is due to their physical property resemblance to living tissue. The temperature or pH sensitive hydrogels could be used in site-specific delivery of drugs to diseased lesions and have been prepared for low molecular weight and protein drug delivery.

Because of environmental concerns, the commercial utilization of biological polymers has become an active research area during past decades (*21-23*). Biopolymers have potential advantages compared with synthetic petroleum polymers owing to their biodegradable properties and, in many cases, lower cost. Vegetable oils are non-toxic renewable resources. It is well known that soybean oil is mainly a mixture of triacylglycerides which are esters of glycerol with various saturated and unsaturated fatty acids. These double bonds in unsaturated fatty acids could be converted into the more reactive oxirane moiety by reaction with peracids or peroxides. Epoxidized soybean oil (ESO), the chemical structure shown in Scheme 1, would be polymerized by ring-opening polymerization (*24*). The formed polymer can be hydrolyzed to lead to the polymers occupying the polyether backbone with alkyl and carboxylic groups, which are useful in the hydrogel formulation.

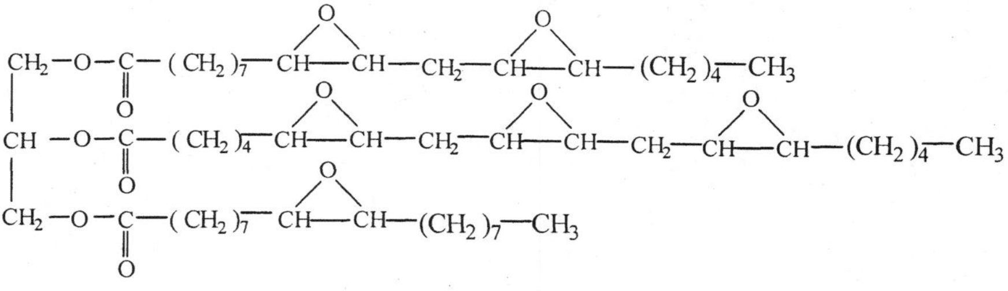

Scheme 1. Chemical structure of ESO.

Molecular Design

Polymeric surfactant, Pluronic series, is the copolymerization of ethylene oxide and propylene oxide. Propylene oxide is hydrophobic and ethylene oxide is hydrophilic, as shown in Scheme 2. By controlling the molecular weight and propylene oxide content, the properties of the surfactant can be changed for a specific application or formulation. Based on the idea of combining hydrophilic and hydrophobic block copolymers, Soy-based polymer from ESO would be designed as structure shown in Scheme 3. The polymer chain is constituted of polyether backbone, attached with a hydrophobic alkyl group and a hydrophilic carboxylic group at each repeat unit. The synthesis of soy-based polymer from ESO, and then hydrolyzed with sodium hydroxide to form the designed structure of polymer (HPESO) shown in Scheme 4, below:

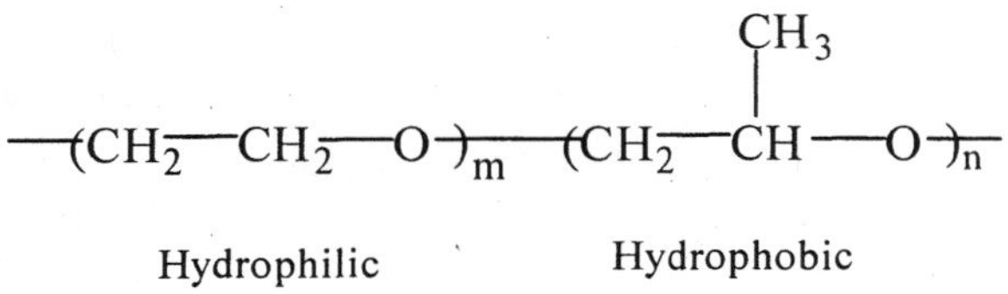

$$-(CH_2-CH_2-O)_m \quad (CH_2-CH-O)_n$$
$$\overset{CH_3}{|}$$

Hydrophilic   Hydrophobic

Scheme 2. Molecular structure of poly(ethylene oxide) (PEO) and poly(propylene oxide) (PPO).

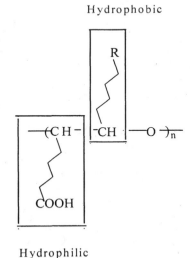

Hydrophobic

Scheme 3. Diagram of designed soy-based molecular structure.

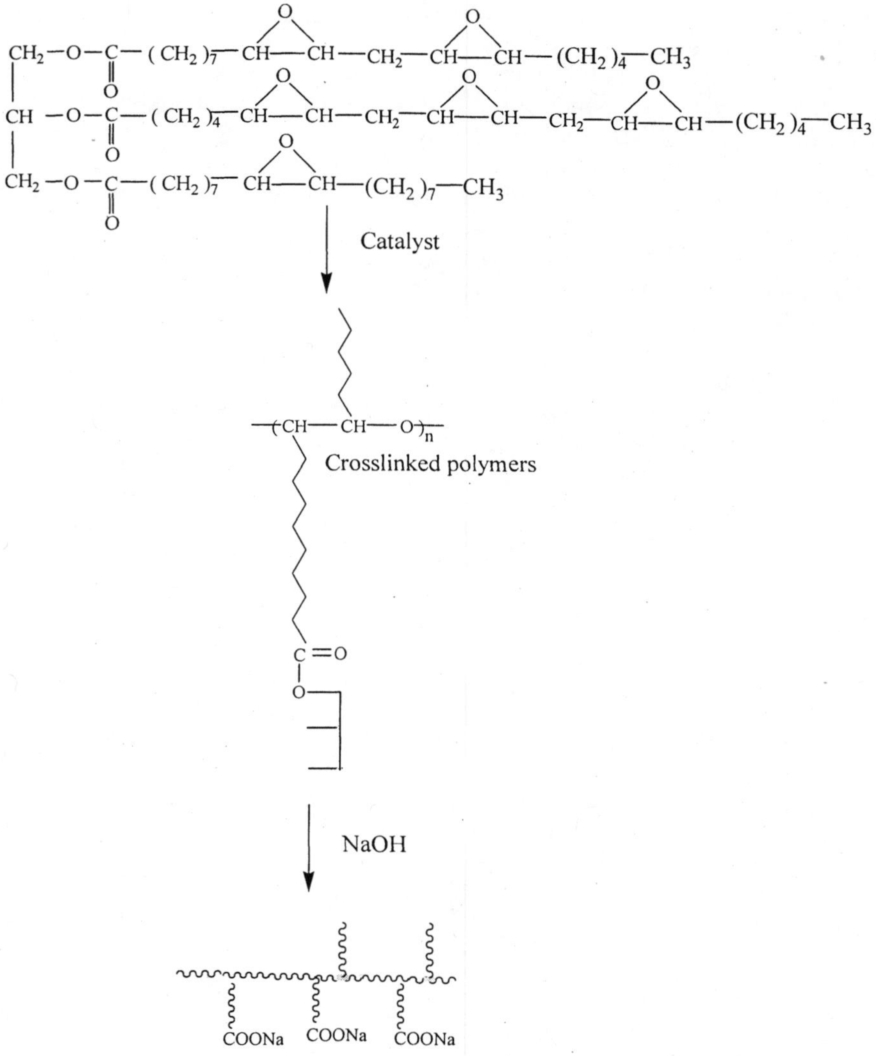

$CH_2-O-C-(CH_2)_7-CH\overset{O}{\overset{\triangle}{-}}CH-CH_2-CH\overset{O}{\overset{\triangle}{-}}CH-(CH_2)_4-CH_3$

$CH-O-C-(CH_2)_4-CH\overset{O}{\overset{\triangle}{-}}CH-CH_2-CH\overset{O}{\overset{\triangle}{-}}CH-CH_2-CH\overset{O}{\overset{\triangle}{-}}CH-(CH_2)_4-CH_3$

$CH_2-O-C-(CH_2)_7-CH\overset{O}{\overset{\triangle}{-}}CH-(CH_2)_7-CH_3$

Catalyst

$-(CH-CH-O)_n-$

Crosslinked polymers

$C=O$

NaOH

COONa   COONa   COONa

Scheme 4. Synthesis of HPESO polymer.

Materials

ESO was obtained from Alf Atochem Inc. (Philadelphia, PA) and used as received. Boron trifluoride diethyl etherate, $(C_2H_5)_2$ O·BF_3, purified and redistilled, was provided by Aldrich Chemical Inc. (Milwaukee, WI). Methylene chloride was purchased from Fisher Scientific (Fair Lawn, NJ).

The catalytic ring opening polymerization of ESO initiated by $(C_2H_5)_2$ O·BF_3 is carried out in methylenechloride at low temperature (0-50°C), resulting in the formation of polymers in high yield. The reaction using ESO is illustrated in reaction scheme 3, below. The proposed reaction can be illustrated in Scheme 3.

ESO Ring-Opening Polymerization in Methylene Chloride

Typically, in a 500 ml flask fitted with a mechanical stirrer, condenser, thermometer, nitrogen line and dropping funnel, 30.0 g ESO and 300 ml methylene chloride were added. Temperature was cooled to 0°C by an ice bath. 0.396g BF_3 (diethyl etherate) was added dropwise in 2 minutes. The system was maintained for 3 h. The methylene chloride was removed using a rotary evaporator and dried under vacuum at 70°C to a constant weight. 29.8g of polymer was obtained (PESO).

Hydrolysis of PESO Using NaOH

PESO (2.5g) in 50 ml 0.4N NaOH was refluxed for 24 h. Then the solution was filtered through filter paper and cooled to room temperature. The resulting gel was precipitated with 80 ml 1.0 N HCl, followed by washing with water several times, and then with 10% (v/v) acetic acid twice. The resulting polymer was dried in the oven at 80°C overnight and then further dried under vacuum at 70°C to a constant weight (2.1g). The product was referred to as HPESO.

FT-IR Analysis

FTIR spectra were recorded on a Therma Nicolet 470 FT-IR system (Madison, WI) in a scanning range of 650-4000 cm^{-1} for 32 scans at a spectra resolution of 4 cm^{-1}. Spectra of PESO and HPESO were run as KBr pellets and spectrum of ESO was run as neat liquid. Infrared spectrum of ESO shown in Figure 1(a) can be compared to the spectrum of PESO, shown in Figure 1(b). As shown, the characteristic oxirane absorption at 823.3 cm^{-1} in ESO is not present in PESO due to the ring-opening polymerization. The IR spectrum of PESO in

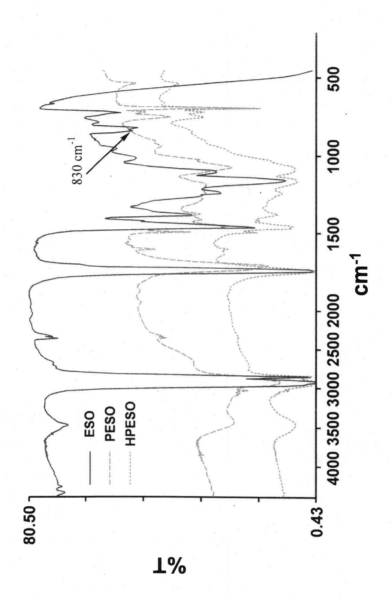

Figure 1(c) obtained after hydrolysis of PESO by NaOH showed a shift in the ester carbonyl band to 1718.9 cm^{-1} from 1734.5 cm^{-1} when compared to PESO due to strong H-bonding of the carboxylic acids as dimers. Hydrogen bonding and resonance weaken the C=O bond, resulting in absorption at a lower frequency.

NMR Analysis

Solid ^{13}C NMR spectra were recorded using a Bruker ARX-300 for PESO samples because they did not dissolve in solvents. ^{1}H and ^{13}C NMR spectra for HPESO samples were recorded quantitatively using a Bruker ARX-400 spectrameter (Bruker, Rheinstetten, Germany) at an observing frequency of 400 and 100 MHz respectively on a 5 mm dual probe. In solid ^{13}C NMR spectrum of PESO, a peak at 75 ppm is due to carbons of the –CH$_2$-CH-CH$_2$- glycerol backbone was observed. Presence of peak at 175 ppm is due to carbonyl carbon of triacylglycerol. In addition, the disappearance of epoxy carbon peaks in the range of 54-57 ppm is observed due to ring-opening polymerization.

Figure 2 shows ^{1}H NMR spectrum of ESO and HPESO. The peak assignments were done using DEPT 135 and COSY 45 NMR experiments. The epoxy protons are observed in the δ 3.0-3.2 ppm region. Methine proton of –CH$_2$-CH-CH$_2$- backbone at δ 5.1-5.3ppm, methylene proton of –CH$_2$ –CH-CH$_2$- backbone at δ 4.0-4.4 ppm, CH$_2$ proton adjacent to two epoxy group at δ 2.8-3.0 ppm –CH- protons of epoxy ring at δ 3.0-3.2 ppm, α-CH$_2$ to >C=O at δ 2.2-2.4 ppm, α-CH$_2$ to poxy group at δ 1.7-1.9 ppm, β- CH$_2$ to >C=O at δ 1.55-1.7 ppm, β- CH$_2$ to epoxy group at δ 1.4-1.55 ppm, saturated methylene groups at δ 1.1-1.4 ppm and terminal –CH$_3$ groups at δ 0.8-1.0 ppm region. ^{1}H NMR spectrum of HPESO indicates disappearance of epoxy carbon peaks in the range of 3.0-3.2 ppm region. In ^{13}C NMR spectrum of ESO, peaks at 54-57 ppm are assigned to epoxy carbons. The presence of ^{13}C NMR peak at 173.1 ppm is due to carbonyl carbon of triacylglycerol. Peaks at 68.9 ppm and 62 ppm, respectively, assign for CH and CH$_2$ carbons of the –CH$_2$-CH-CH$_2$- glycerol backbone. ^{13}C NMR spectrum of HPESO indicates disappearance of epoxy carbon peaks in the range of 54-57 ppm. Peaks at 68.9 ppm and 62 ppm assigned for CH and CH$_2$ carbons of –CH$_2$-CH-CH$_2$- glycerol backbone also disappeared due to hydrolysis of glycerol.

HPLC Analysis

GPC profiles were obtained on a PL-GPC 120 high temperature chromatography (Polymer Laboratories, Ltd. Amherst, MA) equipped with in-built of differential refractometer detector and an autosampler using polymer

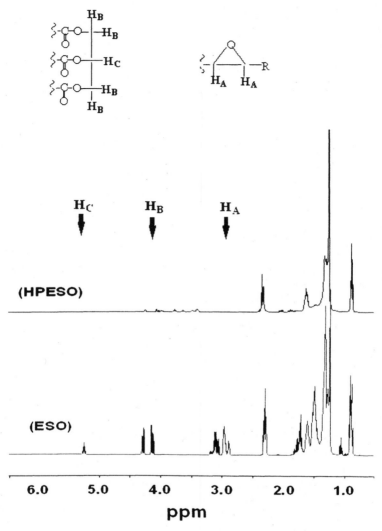

Figure 2. ¹H NMR spectra of ESO, PESO and HPESO.

styrene standard with molecular weights of 1700, 2450, 5050, 7000, 9200 and 10665 g/mole (Polymer Laboratories, Ltd. Amherst, MA) for calibration. HPESO samples for GPC analysis are prepared according to the method described above. HPESO sample with molecular weight, Mw=5866 was selected for viscoelastic properties and controlled drug release system study.

Viscoelastic Properties of HPESO Hydrogel

Because soy-based products contain fatty acid residues that can be readily attacked by lipase-secreting bacteria, the family of the materials is generally regarded as biodegradable. In order to further explore its usage and application, its rheological properties need to be examined. The viscoelastic properties of HPESO hydrogel were investigated. The rheological properties of this novel biological hydrogel indicated that it possessed potential applications for bioengineering of drug delivery and tissue engineering.

Preparation of Hydrogel

HPESO was fully neutralized by NaOH aqueous solution based on the ion-exchange value, and then diluted with phosphate buffer at certain pH into the desire concentration stirred with a magnetic bar for 6 hours at 70°C in a water bath. When the solution was cooled to room temperature, a gel was form. To heat the gel into 55°C, it became solution again and was placed on the plate fixture of the rheometer and the measurements were taken directly.

Measurements

A strain-controlled Rheometric ARES rheometer (TA Instruments, New Castle, DE) was used to perform the rheology studies. A 50-mm diameter cone and plate geometry was adopted. The cone angle was 0.04 radians. The temperature was controlled at 55±0.1°C or 25±0.1°C by a water circulation system. A steady shear experiment was performed only at 55±0.1°C. Linear viscoelastic measurements were conducted for HPESO hydrogels. Five concentrations of HPESO samples from 1% to 5% (wt.-%) were investigated. At 55°C, HPESO sample behaved as a perfect viscous liquid with a viscosity close to water. No elasticity (G') could be detected at 55°C. Its viscosity was about 1.1 cP and nearly independent of shear rate at 55°C. These results indicated that at 55°C, HPESO displayed Newtonian fluid and viscous liquid behavior at measured concentrations.

In order to evaluate the rheological behavior of the HPESO hydrogel formation, a series of viscoelastic measurements were conducted. At first, the thermal dependence of gelation for HPESO was examined by cycling a HPESO gel through a series of test of dynamic temperature steps ranging from 55°C to 25°C at pH 8.0. Figure 3 displays the change of the elastic modulus (G') versus temperature for a 3% HPESO sample. The solid curve in Figure 3 shows the data obtained from 55°C to 25°C during the cooling portion of the cycle; the dashed line shows the heating from 25°C back to 55°C. Although the kinetics of the de-

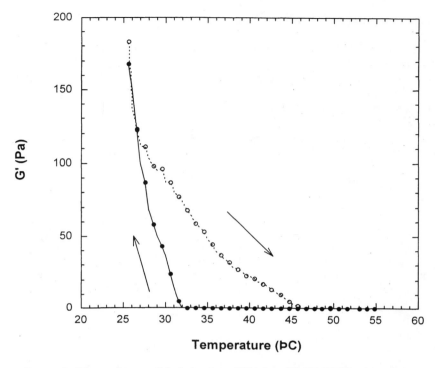

Figure 3. Thermal reversible behavior of 3% (wt.-%) HPESO hydrogel with 1 rad/s frequency. Solid line and filled symbols: storage modulus (G') versus temperature during cooling from 55°C to 25°C. Dashed line and opened symbols: storage modulus (G') versus temperature during heating from 25°C to 55°C.

gelation and gelation was different due to the various rates of association (assembly) and dissociation (disassembly) of the molecules, this thermal sol-gel procedure was completely reversible and repeatable. The same tend was also found for the other concentrations of 2% to 5% HPESO. Because 1% HPESO is a too diluted solution, it could not form a gel but still exhibited a weak viscoelastic behavior at 25°C. The pH range for the HPESO to form a gel is between 7.5 and 14, which is from the physiological level to the basic level.

To monitor the gelation process, a time sweep measurement was conducted at 25°C with 0.05% of strain amplitude and 1 rad/s frequency. At 25°C, 3% and 5% HPESO immediately exhibited viscoelastic behavior (Figure 4), which implied that the molecules cross-linking and association. The storage or elastic modulus (G') jumped to 90 Pa (3% HPESO) and 200 Pa (5% HPESO) respectively within a few seconds. A plateau of about 1200 Pa was reached after 3 hours for 5% HPESO, and a plateau of around 260 Pa after 8 hours for 3%

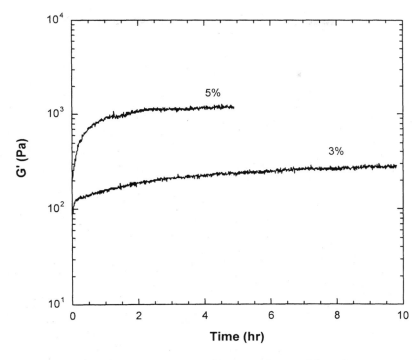

Figure 4. Storage modulus (G') as a function of time of the gelation procedure for 3% and 5% (wt.-%) HPESO hydrogel with 1 rad/s and 0.05% strain at 25°C.

HPESO. The gelation process and stable value of G' take at least a few hours, which means that the assembly and cross-linking of the HPESO molecules require a fair amount of time to achieve full cure.

At the equilibrium at 25°C, the $\geq 2\%$ concentrations of HPESO exhibited viscoelastic solid properties (Figure 5). For example, the storage moduli (G') curve for the 5% HPESO had a plateau of 1200 Pa over three frequency decades. The loss moduli (G") also had a plateau in the order of 160 Pa. The phase shifts were 5.5° – 13.8°. The shapes of G' and G" curves were very similar to those of gels (25). 1% HPESO displayed viscoelastic liquid behavior (Figure 5). Its G" was greater than G' over the most range of the measured frequency, and the phase shifts were 20° – 80°. The viscoelastic properties of the HPESO were strongly dependent on the concentrations (Figure 4 & 5). The higher concentration, the greater values of G' would be (Figure 5).

The function and behavior of the HPESO hydrogel suggest that this biomaterial be a candidate for application in bioengineering of drug delivery and tissue engineering.

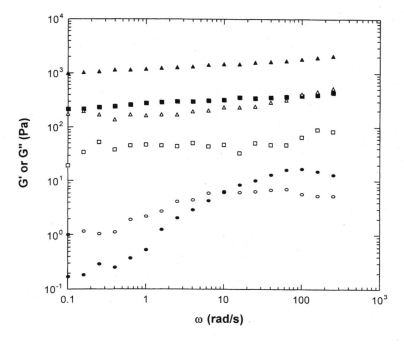

Figure 5. Linear viscoelastic properties of frequency sweep experiment for HPESO. Storage modulus (G') or loss modulus (G") as function of frequency at 25°C with 0.05% strain. Filled symbols: G', opened symbols: G". Circles: 1% (wt.-%) HPESO, squares: 3% (wt.-%) HPESO, triangles: 5% (wt.-%) HPESO.

Application for Controlled Drug Release

Drug resistance is common cause of treatment failure in the management of several cancer types including breast cancer (*26,27*). There are multiple mechanisms that underlie clinical drug resistance, including noncellular and cellular mechanisms. One of the most studied cellular mechanisms is the classical form of multidrug resistance (MDR), which is usually mediated by the overexpression of P-glycoprotein (P-gp) and other membrane transporters (*28*). To overcome drug resistance and to improve the effectiveness and safety of cancer chemotherapy, new drug delivery systems such as microspheres, nanoparticles, and liposomes have been studied (*29-42*). In comparison to the conventionally used drug solutions, these systems generally exhibit lower toxicity and thus allow higher doses of drugs to be safely administered. This is of clinical significance because many cytotoxic agents have low therapeutic indices.

Solid lipid nanoparticles (SLNs) are colloidal drug carriers with great potential to improve chemotherapy of MDR cancer, yet this area remains largely unexplored. Wu (*43*) have developed polymer-lipid hybrid formulation of SLNs consisting of dextra sulfate, an anionic polymer that can form complexes with a range of cationic drugs. Increased drug loading has been obtained for cationic drugs including doxorubicin (Dox), a broad-spectrum cytotoxic anticancer drug. Up to 90% of the loaded Dox was released in 18 h by using a combination of ion-exchange and diffusion-controlled mechanisms. Their promising results lead to further explore this soy-based polymer-lipid nanoparticle system suit the purpose of MDR cancer treatment. This new soy-based anionic polymer could improve the formulation of the SLN-containing Dox (Dox-SLN) to achieve better particle morphology and drug loading capacity. The effect of the polymer-lipid hybrid SLN system on the cytoxicity of Dox against human MDR breast cancer cells (MDA435/LCC6/MDR1) was investigated. This cell line was chosen because it highly expresses a classical membrane transporter, P-gp-MDR1 by trandusction of MDA435/LCC6 cell line with a retroviral vector directing the constitutive expression of the MDR1 cDNA (*44*).

Preparation of SLNs and Drug-Polymer Aggregates Containing Dox

Dox-SLNs were prepared similarly to a method reported by Wu (*43*), with the addition of an ultrasonication step to improve lipid dispersion (*45*). Briefly, a mixture of 50 mg stearic acid and 0.45 mL of aqueous solution containing 0.5 to 5 mg Dox and Pluronic F68 (2.5% w/v) was warmed to 75°C in a water bath. HPESO polymer solution preheated to 72-75°C was added to the mixture (Dox/polymer w/w ratio =2:1). The final mixture formed was stirred for 10 min and then ultrasonicated for 3 min to form submicrometer-sized lipid emulsion. Dox-SLNs were formed by dispersing 1 volume of the emulsion in 4 to 9 volumes of water at 4°C. Blank SLNs were prepared in the same manner except Dox was omitted.

Dox-HPESO aggregates were prepared by magnetically stirring the mixture of aqueous solutions of Dox and HPESO in 2:1 w/w (Dox/polymer) ratio in the dark until a suspension of fine aggregates was formed. The suspension was centrifuged at 1.000 × g for 2 min. The supernatant containing the uncomplexed drugs were drawn out and the pellet was washed with ice-cold distilled water. The washout liquid was pooled with the supernatant and the amount of uncomplexed Dox was measured by spectrophotometry from which the complexed drug was calculated. It was estimated that 67% of Dox was complexed with the polymer (data not shown). Dox-HPESO pellets were resuspended in 2.5% Pluronic F68 and diluted to the desired concentrations for the cytotoxicity tests.

The TEM image of a Dox-SLN sample shown in Figure 6 illustrates that the particles are generally spherical in shape and the majority of the particles are in the range 50 to 200 nm in diameter. The negative surface charge is indicative of the exposure of the anionic HPESO chains on the surface. The use of HPESO that contains carboxylic groups has improved the quality of the SLN. As compared to the formulation consisting of a more hydrophilic polymer, dextran sulfate (DS) (*43*), the HPESO-containing SLNs were more uniform and more spherical. Unlike the SLNs made from DS which formed large agglomerates when the dug loading was over 4%, HPESO-containing SLN showed much less aggregation at Dox loading up to 6%. This property enables the preparation of Dox-SLN with higher drug loading than the DS-based system.

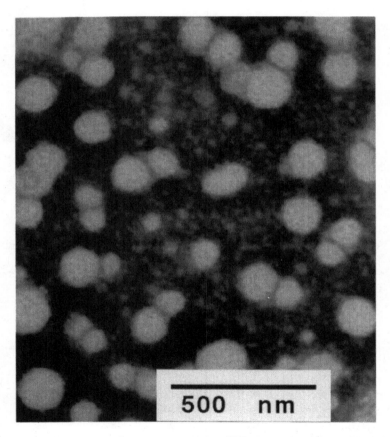

Figure 6. Transmission electron microscopy (TEM) image of Dox-SLN (40,000 magnification). The particles were negatively stained with aqueous solution of phosphotungstic acid (PTA) 5 min prior to imaging. The sample used was loaded with 3.8% w/w Dox.

Drug Loading and Encapsulation Efficiency

Figure 7 presents the percent drug loaded per weight of nanoparticle (% DL) and encapsulation efficiency of Dox (EE) in Dox-SLN as a function of Dox added per 100 mg of the lipid. As the amount of Dox is increased from 2 to 10 mg, % DL progressively increases from less than 2 to above 6% w/w. EE remains between 70 to 80% up to 6 mg of Dox added and only declines slightly to about 65% up to 10 mg of Dox added. In the absence of the anionic polymer, the EE values were 38 and 28%, respectively (n = 2 in each case, data not shown), for SLNs prepared with 2 and 5 mg of Dox per 100 mg of lipid.

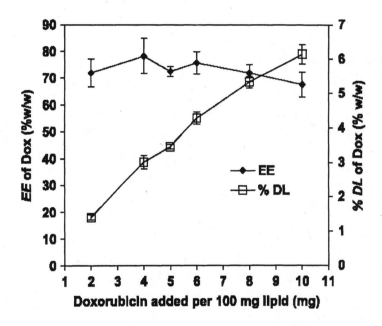

Figure 7. Effect of the amount of drug fed in Dox-SLN preparation on drug encapsulation efficiency (EE) and percent of drug loading (% DL). Results were expressed in terms of Dox loaded per weight of nanoparticles in Dox-SLN as a function of amount of Dox in the feed per 100 mg of lipid. Each value represents mean ± SD (n ≥ 3).

Drug Release Kinetics of Dox-SLN and Dox-HPESO

A typical drug release profile of Dox-SLN with 3.5% drug loading is shown in Figure 8. Half of the drug loaded in the nanoparticles is released in about 2 to 4 h, and an additional 10% of the loaded drug is gradually released in another 12

h. The total amounts of drug released from the nanoparticles after 72 h and 2 weeks (not shown in the graph) were 66.2 ± 4.1 and 70.6 ± 4.5% (n = 3, mean ± SD), respectively. For comparison, the drug release profile from Dox-HPESO was also obtained. A quick release was observed with approximate 35% of Dox being released in 15 min and a plateau at about 40% reached after 1 h. Without a lipid barrier, the drug molecules that are bound to the surface of the polymer aggregates by ionic complexation can be rapidly released in the presence of counterions in the buffer. Because the aggregates are large (several micrometers to millimeters) and the polymers are densely packed, it is likely that a considerable portion of drug locked up in the inner cores of these aggregates is hardly available for release and contributes to the later slow-release phase.

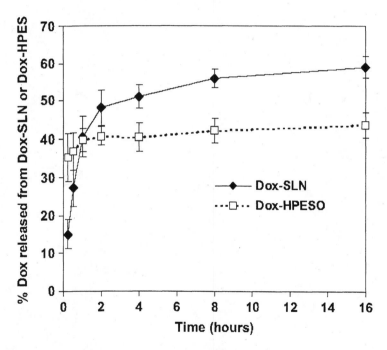

Figure 8. Drug release profiles of Dox from Dox-SLN (average drug loading of three different batches = 3.5% w/w) or Dox-HPESO (containing 3.5 mg Dox in 5 mL) in PBS (pH 7.4, m = 0.15 M) at 37°C. Not shown in the Dox-SLN profile are the values of percent Dox released after 72 h and 2 weeks, which were measured as 66.2 ± 4.1 and 70.6 ± 4.5%, respectively. Each value represents mean ± SD of the measurements obtained in three separate experiments.

Influence of Dox-SLNs and Various Treatments on Cell Membrane Integrity

As shown in Figure 9A and B, Dox-SLN and the components of the formulation (HPESO polymer, Dox, Dox-HPESO, blank SLN) do not significantly reduce the normalized membrane integrity of either wild-type (Fig. 9A) or drug-resistant (Fig. 9B) cancer cells for exposures of 4 h or less to below 1. Decreases in normalized membrane integrity are seen only in the treatments that contain Dox (Dox, Dox-HPESO, and Dox-SLN) for longer periods of time (8 and 24 h). These decreases are larger in the wild-type cells (Fig. 9A) than in the MDR cells (Fig. 9B). It is noticed that a significant decrease in membrane integrity happens only after 24 h of Dox-containing treatment (Dox solution, Dox-SLN, Dox-HPESO). Because both blank SLN and HPESO are nontoxic to the cells (46), this toxicity was likely contributed by Dox in these formulations. Among the three Dox-containing formulations, Dox-SLNs are comparatively most toxic, even more toxic than Dox in wild-type cells. In addition to more Dox release from the nanoparticles, the extended duration of treatment (24 h instead of 4 h used in clonogenic assay) could possibly allow more cell-particle interactions. The use of HPESO that contains carboxylic groups has improved the quality of the SLN. As compared to the formulation consisting of a more hydrophilic polymer, dextran sulfate (DS) (43), the HPESO containing SLNs was more uniform and more spherical. Unlike the SLNs made from DS which formed large agglomerates when the dug loading was over 4%, HPESO containing SLN showed much less aggregation at Dox loading up to 6%. This property enables the preparation of Dox-SLN with higher drug loading than the DS-based system.

Figure 10A and B compares the tumour cell toxicity of 4 h exposures to free DOX solution, Dox-HPESO polymer aggregates (Dox-polymer aggregates), Dox-SLN, and Dox plus blank SLN (blank SLN added separately to free Dox solution) on wild-type and MDR breast cancer cell lines, respectively. As expected, all four types of Dox-containing treatments are more cytotoxic to wild-type cells than to MDR cells; the normalized plate efficiency, which was determined by dividing the plating efficiency (PE) of treated cells by PE of control (untreated cells), of MDR cells is at least one log scale higher than the wild-type cells receiving the same treatment type. Note that the range of Dox concentrations in Figure 10A (0.5-5 mg/mL) is lower than those in Figure 10B (1-10 mg/mL). As demonstrated in Figure 10A, Dox-polymer aggregates were more cytotoxic than Dox solution, whereas the cytotoxicity of Dox-SLN was similar to that of Dox solution. A different pattern is observed in the experiments with MDR cells (Figure 10B). Dox-SLN treatment is significantly more toxic toward MDR cells than other treatments with over 8-fold cell kill being achieved in the higher concentration range. On the other hand, Dox-polymer aggregates exhibit no difference in cytotoxicity against MDR cells as compared to Dox

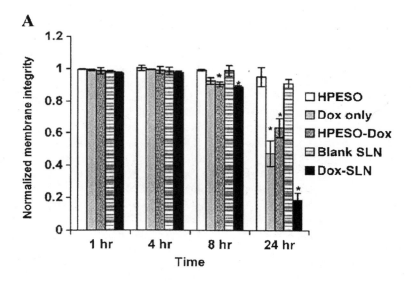

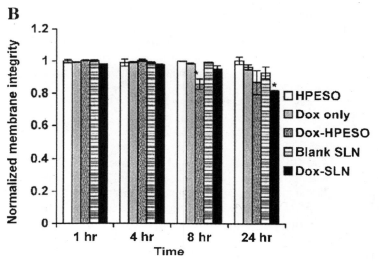

Figure 9. Results of trypan blue exclusion assay experiments showing the effects of HPESO solution, Dox-HPESO aggregates, blank SLN, and Dox-SLN on cell membrane integrity of (A) MDA435/LCC6/WT cells and (B) MDA435/LCC6/MDR1 cells. Cells were treated for 1, 4, 8, and 24 h. The total concentrations of Dox, HPESO polymer, and stearic acid used were 10, 5, and 200 2g/mL, respectively. Results were normalized against the control (drug-free PBS). Each value represents mean ± SD of the measurements obtained in three separate experiments (n = 2 in each experiment).

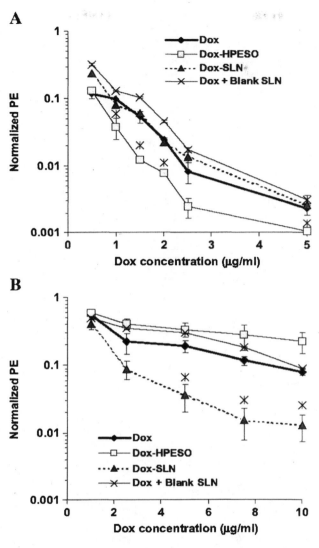

Figure 10. Clonogenic assay experiments for the toxicity of Dox in (A) MDA435/LCC6/WT and (B) MDA435/LCC6/MDR1 cell lines. The normalized plating efficiencies (normalized PE) after 4 h exposure to Dox solution, Dox-HPESO aggregates, Dox-SLN, or Dox solution + blank SLN are shown. In treatments that included HPESO polymer (Dox-polymer aggregates and Dox-SLN) or lipid (Dox-SLN and Dox + blank SLN), the w/w ratios of polymer to Dox and lipid to Dox were equally set at 1:0.48 and 1:20, respectively, at all Dox concentrations. Results are expressed as mean ± SD of the measurements obtained in three separate experiments (n = 6 in each experiment).

solution. As seen in Figure 10A and B, the addition of blank SLN to fresh Dox solution does not increase the cytotoxicity toward either cell type.

The treatment of the MDR cells with Dox-SLN resulted in over 8-fold increase in cell kill when compared to Dox solution treatment at equivalent doses. These data suggest that the new polymer-lipid nanoparticle system may offer great potential to deliver Dox effectively for the treatment of MDR breast cancer.

References

1. Bae, Y. H.; Huh, K. M.; Kim, Y.; Park, K. H. *J. Controlled Release* **2000**, *64*, 3-13.
2. Gan, Z. H.; Jim, T. F.; Li, M.; Yuer, Z.; Wang, S. G.; Wu, C. *Macromolecules* **1999**, *32*, 590-594.
3. Ge, H. X.; Hu, Y.; Jiang, X. Q. *J. Pharm. Sci.* **2002**, *91*, 1463-1473.
4. Kosita, M. J.; Bohne, C.; Alexandridis, P.; Hatton, T. A.; Holzwarth, J. F. *Macromolecules* **1999**, *32*, 5539-5551.
5. Lee, S. H.; Kim, S. H.; Kim, Y. H. *Macromol. Res.* **2002**, *10*, 85-90.
6. Wu, C.; Liu, T.; Chu, B.; Schneideer, D. K.; Graziano, V. *Macromolecules* **1997**, *30*, 4574-4583.
7. Yekta, A.; Xu, B.; Duhamel, J.; Adiwidjaja, H.; Winnik, M. A. *Macromolecules* **1995**, *28*, 956-966.
8. Yuan, M. L.; Wang, Y. H.; Li, X. H.; Xiong, C. D.; Deng, X. M. *Macromolecules* **2000**, *33*, 1613-1617.
9. Zhao, Y.; Liang, H. J.; Wang, S. G.; Wu, C. *J. Phys. Chem. B* **2001**, *105*, 848-851.
10. Rosler, A.; Wandermeulen, G. W.; Klok, H. A. *Adv. Drug Delivery Rev.* **2001**, *53*, 95-108.
11. Hurter, P. N.; Alexandritis, P.; Hatton, T. A. In *Solubilization in Surfactant Aggresgates*; Christian, S. D.; Scamehorn, J. F., Eds.; Marcel Dekker: New York, 1995.
12. Ha, J. C.; Kim, S. Y.; Lee, Y. M. *J. Controlled Release* **1999**, *62*, 381-392.
13. Kim, S. Y.; Ha, J. C.; Lee, Y. M. *J. Controlled Release* **2000**, *65*, 345-358.
14. Bromberg, L. *Ind. Eng. Chem. Res.* **1998**, *37*, 4267-4274.
15. Bromberg, L. *Langmuir* **1998**, *14*, 5806-5812.
16. Bromberg, L. *Macromolecules* **1998**, *31*, 6148-6156.
17. Xiong, X. Y.; Tam, K. C.; Gan, L. H. *Macromolecules* **2003**, *36*, 9979-9985.
18. Chen, G.; Hoffman, A. S.; Ron, E. S. *Proc. Intern. Symp. Control. Rel. Bioact. Mater.* **1995**, *22*, 167.
19. Ron, E. S.; Roos, E. J.; Staples, A. K.; Schiller, M. E.; Bromberg, L. *Proc. Intern. Symp. Control. Rel. Bioact. Mater.* **1996**, *23*, 128.

20. Ron, E. S.; Roos, E. J.; Staples, A. K.; Schiller, M. E.; Bromberg, L.; *Pharm. Res.* **1996** (Suppl.), *13*, S299.

21. Kabanov, A. A.; Alakhov, V. Yu. In *Amphilic Block Copolymers: Self Assembly and Applications*; Alexandrities, P.; Lindman, B., Eds.; Elsevier: Amsterdam, 1997.

22. Hurter, P. N.; Alexandritis, P.; Hatton, T. A. In *Solubilization in Surfactant Aggresgates*; Christian, S.D.; Scamehorn, J. F., Eds., Marcel Dekker: New York, 1995.

23. Wanka, G.; Hoffman, H.; Ulright, W. *Macromolecules*, **1994**, *27*, 4145: Malmsten, M. Lindman, B., *Macromolecules*, **1992**, *25*, 5440.

24. Liu, Z.; Erhan, S. Soy-based Thermosensitive Hydrogels for Controlled Release Systems, U.S. Patent, pending.

25. Clark, A. H., In *Physical Chemistry of Foods;* Schwartzberg, H. G.; Hartel, R. W., Eds.; Marcel Dekker, Inc.: New York, 1992; pp 263-305.

26. Lonning, P. E. *Lancet Oncol.* **2003**, *4*, 177-185.

27. Early Breast Cancer Trialists' Collaborative Group. *Lancet* **1998**, *352*, 930-942.

28. Gottesman, M. M. *Annu. Rev. Med.* **2002**, *53*, 615-627.

29. Liu, Z.; Wu, X. Y.; Bendayan, R. *J. Pharm. Sci.* **1999**, *88*, 412-418.

30. Cheung, R. Y.; Rauth, A. M.; Wu, X. Y. *Anti-Cancer Drugs* **2005**, *16*(4), 423-433.

31. Cheung, R.; Ying, Y.; Rauth, A. M.; Marcon, N.; Wu, X. Y. *Biomaterials* **2005**, *26*, 5375-5385.

32. Liu, Z.; Cheung, R.; Wu, X. Y.; Ballinger, J. R.; Bendayan, R.; Rauth, M. *J. Control. Release* **2001**, *77*, 213-224.

33. Steiniger, S. C. J.; Kreuter, J.; Khalansky, A. S.; Skidan, I. N.; Bobruskin, A. I.; Smirnova, Z. S.; Severin, S. E.; Uhl, R.; Kock, M.; Geiger, K. D.; Gelperina, S. E. *Int. J. Cancer* **2004**, *109*, 759-767.

34. Gelperina, S. E.; Khalansky, A. S.; Skidan, I. N.; Smirnova, Z. S.; Bobruskin, A. I.; Severin, S. E.; Turowski, B.; Zanella, F. E.; Kreuter, J. *Toxicol. Lett.* **2002**, *126*, 131-141.

35. Koukourakis, M. I.; Koukouraki, S.; Fezoulidis, I.; Kurias, G.; Archimandritis, S.; Karkavitsas, N. *Br. J. Cancer* **2000**, *83*, 1281-1286.

36. Massing, U.; Fuxius, S. *Drug Resist. Updat.* **2000**, *3*, 171-177.

37. Liu, Z.; Ballinger, J. R.; Rauth, A. M.; Bendayan, R.; Wu, X. Y. *J. Pharm. Pharmacol.* **2003**, *55*, 1063-1073.

38. Liu, Z.; Bendayan, R.; Wu, X. Y. *J. Pharm. Pharmacol.* **2001**, *53*, 1-12.

39. Liu, Z.; Wu, X. Y.; Bendayan, R. *J. Pharm. Sci.* **1999**, *88*, 412-418.

40. Huwyler, J.; Cerletti, A.; Fricker, G.; Eberle, A. N.; Drewe, J. *J. Drug Target.* **2002**, *10*, 73-79.

41. Mamot, C.; Drummond, D. C.; Hong, K.; Kirpotin, D. B.; Park, J. W. *Drug Resist. Updat.* **2003**, *6*, 271-279.

42. Soma, C. E.; Dubernet, C.; Bentolila, D.; Benita, S.; Couvreur, P. *Biomaterials* **2000**, *21*, 1-7.

43. Wong, H. L.; Bendayan, R.; Rauth, A. M.; Wu, X. Y. *J. Pharm. Sci.* **2004**, *93*, 1993-2004.

44. Leonessa, F.; Green, D.; Licht, T.; Wright, A.; Wingate-Legette, K.; Lippman, J.; Gottesman, M. M.; Clarke, R. *Br. J. Cancer* **1996**, *73*, 154-161.

45. Hou, D. Z.; Xie, C. S.; Huang, K. J.; Zhu, C. H. *Biomaterials* **2003**, *24*, 1781-1785.

46. Wong, H. L.; Rauth, A. M.; Bendayan, R.; Manias, J. L.; Ramaswamy, M.; Liu, Z. S.; Erhan, S. Z.; Wu, X. Y. *Pharmaceutical Research* **2006**, *23*(7), 1574.

Chapter 3

Application of Lipases to Develop Novel Acyltrnsferase Substrates

Thomas Mckeon, Charlotta Turner, Sung-Tae Kang, Xiaohua He, Grace Chen, and Jiann-Tsyh Lin

Western, Regional Research Center, Agricultural Research Service, U.S. Department of Agriculture, 800 Buchanan Street, Albany, CA 94710

Introduction

The castor plant (*Ricinus communis*) produces seed containing a unique oil that has numerous non-food applications (*1*). Up to 90% of the fatty acid content of the castor oil is ricinoleate (12-hydroxy oleate). Table 1 illustrates the fact that ricinoleate and its derivatives have many important industrial uses, such as lubricants, coatings, plastics and fungicides. However, castor cultivation and processing can expose workers to potent allergens and the castor meal obtained after processing to extract oil also contains the highly toxic seed protein ricin and its less toxic homologue *Ricinus communis* agglutinin (RCA). Because of the potential for expanded use of castor oil in industrial and bio-diesel applications, it is of great interest to develop a safe castor crop, and one logical approach is through genetic engineering a commodity oilseed or a microbe.

Ricinoleate is produced in castor by the action of the oleoyl-12-hydroxylase on oleate in the *sn*-2 position of oleoyl-phosphatidyl choline (*2*). Since transgenic expression of the gene for the hydroxylase produces and oil of no more than 20 % ricinoleate, we have identified other enzymes that appear to underlie the high ricinoleate content of castor oil (*2*).

Table I. Some Products Derived from Castor Oil and Ricinoleic Acid

Lubricants	-Lithium grease
	-Heptanoate esters for engines
Coatings	-Nonyellowing drying oil
	-Low VOC oil-based paints
Surfactants	-Turkey Red Oil
Plasticizers	-Blown oil, for polyamides, rubber
	-Heptanoates, low temperature uses
Cosmetics	-Lipstick
Pharmaceuticals	-Laxative
Polymers	-Polyesters, from sebacic acid
	-Polyamides, Nylon 11, Nylon 6,10
	-Polyurethanes
Perfumes	- 2-octanol, heptanal and undecenal
Deodorant	-Zinc ricinoleate
Fungicides	-Undecenoic Acid and Derivatives

As part of our research approach to identify the enzymes that lead to high ricinoleate in castor oil biosynthesis, we have cloned the cDNA for diacylglycerol acyltransferase (DGAT) from castor seed, one of the enzymes involved in the biosynthesis of triricinolein from 1,2-diricinolein. This enzyme carries out the final step in the biosynthetic pathway of oil in castor seeds (*Ricinus communis L.*), the acylation of 1,2-diricinolein to triricinolein (3), which represents approximately 70% of the TG in castor oil. In order to characterize the activity of the expressed DGAT enzyme, we need 1,2-diricinolein as a substrate, and to better characterize the specificity of castor DGAT, we also need non-native substrates.

The 1,2-diacylglycerols (DG) of hydroxy fatty acids are difficult to synthesize using conventional chemical methodologies, because of the similar reactivity of the hydroxy groups on the fatty acid with those on the glycerol backbone. The ready availability of triricinolein from castor oil provides a suitable starting material, and the choice of lipases provides potential selectivity for the reaction. Both enzymatic and chemical hydrolysis result in the formation of a mixture of 1,2-, 2,3- and 1,3-isomers of DG, in varying proportions. We have used lipases to catalyze hydrolysis and methanolysis of triricinolein to 1,2(2,3)-diricinolein, in order to determine the best conditions for isolating this compound (*4*).

By definition, triacylglycerols carrying unnatural fatty acids do not exist in plant, animal or microbial sources and, so, must be synthesized. The most straightforward method for carrying out such a synthesis is to use enzymatic esterification. Lipase (EC 3.1.1.3) catalytic activity in organic solvents has been studied intensively (*5*). However the application of the solvent free system has been limited because most of the fatty acid reactants have high melting points and the most useful enzymes are not always stable at higher temperatures. We chose to examine the use of a solvent-free system to synthesize the TG of 12-hydroxy stearic acid, a fatty acid that is produced from hydrogenation of ricinoleic acid.

Experimental Procedures

HPLC

HPLCsystem, solvents and sources of chemicals used have been described (4).

Enzymes

Novozyme 435 (immobilized *Candida antarctica* lipase type B, CALB, 10.000 Propyl Laurate Units/g) and Lipozyme RM IM (immobilized *Rhizomucor miehei* lipase, RML, 6.1 BAUN/g) were generous gifts from Novozymes North America (Franklinton, NC). Crude lipases from *Pseudomonas cepacia* (PCL) and *Penicillium roquefortii* (PRL) were kind gifts from Amano Enzymes Inc. (Nagoya, Japan).

Preparation of Triricinolein

Triricinolein is not commercially available, but was produced in our lab by preparative chromatography of castor oil. 0.5 mL of castor oil (0.5 mg/mL in 2-propanol) was injected onto the preparative column. The mobile phase used was a gradient of methanol (A) and 2-propanol (B), and the flow rate was 7 mL/min, eluted with 100% A for 5 min, a linear gradient to A/B (80:20) over 10 min and hold for 5 min. The column was then reconditioned back to 100% A over 5 min and hold for 5 min. Triricinolein, the major peak of castor oil, was collected at a retention time around 10 min. The solvent was removed under nitrogen and mild heating. The purity of the fractions was checked by analytical HPLC.

Immobilization of Lipases

Lipases of *Pseudomonas cepacia* and *Penicillium roquefortii* were immobilized on a polypropylene powder (Accurel MP1000, formerly called EP100) using adsorption as described (4).

Equilibration of Reaction System

Immobilized enzymes, reaction media (hexane and DIPE) and methanol were allowed to equilibrate for at least 48 hours to a certain water activity (a_w) over saturated salt solutions (6). This was achieved in airtight desiccators, in which saturated water solutions of LiCl, $MgNO_3$ and K_2SO_4 gave a_w of 0.11, 0.53 and 0.97, respectively.

Methanolysis Reaction

100 mg of triricinolein was dissolved in 5 mL of reaction media and 0.5 mL of methanol. 100 µL was taken out as a first fraction to find the starting concentration of triricinolein (which was set to 100% yield on the axis of the figures 5-6). Thereafter, 100 mg of immobilized enzyme was added, and the reaction took place in rotating glass tubes at room temperature. 100-µL fractions were taken out at time intervals, the solvent of each collected fraction was removed by nitrogen, and 0.5 mL of 2-propanol was added. The fractions were stored in a freezer at –20°C until HPLC analysis.

HPLC Analysis

Expected products of the lipase reaction, ricinoleic acid, methyl ricinoleate, diricinolein and triricinolein, were analyzed by injection of 20 µL on a RP-HPLC column, using a mobile phase of methanol (A) and methanol/water (90:10) (C). The gradient was as follows: from 100% C to 100% A in 20 min, then hold for 10 min, and a gradient back to 100% C in 5 min, then hold for 5 min. The total run time was 40 min. The flow rate was 1.0 mL/min, and the detection was done at 205 nm. The peaks were quantified using external calibration with methyl ricinoleate at three different concentrations between 0.5 and 5.0 mg/mL.

Optimized Methanolysis Reaction

Enzymes and solvents were pre-equilibrated to a_w 0.53 for at least 48 hours. 100 mg of triricinolein was dissolved in 5 mL of DIPE and 0.5 mL of methanol. 100 mg of PRL was added, and the mixture was allowed to react for 24 hours during stirring at room temperature. The enzyme was removed by filtration at the end of the reaction time. The solvent was removed by nitrogen, and one mL of 2-propanol was added.

Fractionated Collection of Diricinolein

0.5-mL portions were injected onto the preparative column. A mobile phase consisting of methanol (A) and methanol/water (90:10) (C) was used at a flow rate of 7.0 mL/min. The gradient elution was from 100% C to 100% A in 15 min, then held for 15 min, back to 100% C in 5 min, and held for 5 min. The total run time was 40 min. The diricinolein peak was collected at a retention time of 16 min. Anhydrous sodium sulfate was added to the collected fractions, and the tubes were shaken for 60 min. The sodium sulfate was removed by filtration and the solvent was removed by nitrogen under mild heating. The recovered diricinolein oil was stored at -20°C.

NP-HPLC

Peak purity of 1,2(2,3)-isomer *vs.* 1,3-isomer of diricinolein was also determined by NP-HPLC. 20 μL samples in DIPE were injected onto a CN column, using a mobile phase consisting of n-hexane and MTBE (7:3), isocratic run. The total analysis time was 16 minutes. The flow rate was 1.0 mL/min with UV detection at 205 nm. The 1,3- and 1,2(2,3)-isomers eluted two minutes apart, at around 11 and 13 min retention time, respectively.

Cloning and Expression of a cDNA Encoding DGAT from A. thaliana, Columbia (AtDGAT)

A full length AtDGAT cDNA was amplified from an RNA sample extracted from Arabidopsis leaves by RT-PCR. Specific primers were designed based on the sequence information from GenBank (AF 051849): 5'-GAAATGGCGATTTTGGATTCTGCT-3' and 5'- TGACATCGATCCTTTTC-GGTTCAT-3'. The cDNA was cloned into a pYES2.1/V5-His-TOPO vector (Invitrogen, Carlsbad, CA) and verified by complete sequencing in both directions using a Perkin Elmer Big Dye sequencing kit (Perkin Elmer, UK). The recombinant protein was expressed in *Saccharomyces cerevisiae* strain INVSc1 according to the manufacturer's instruction (Invitogen, Carlsbad, CA). Briefly, a single colony containing pYES2.1/V5-His/*AtDGAT* construct was inoculated into medium containing 2% glucose and grown overnight at 30°C with shaking. 2% galactose was added to the medium to induce expression of *AtDGAT* from the *GAL1* promoter. Cells were harvested after induction and the cell pellets were stored at -80°C until ready to use.

Assay for DGAT Activity using Lipase-Catalyzed 1,2(2,3)-Diricinolein as a Substrate

Microsomes were isolated from harvested yeast cells as described (5) and resuspended in 0.1 M Tris-HCl, pH7.0 containing 20% glycerol and kept frozen at -80°C. Protein concentration was determined using the bicinchoninic acid (BCA) protein assay kit (Pierce, Rockford, IL 61105). DGAT assays were performed as described (5). ^{14}C-labelled oleoyl-CoA was synthesized according to McKeon et al. (7). The reaction mixture (100 µl) consisted of 0.1 M Tris-HCl, pH7.0 containing 20% glycerol, microsomes (50 µg of protein), 1,2(2,3)-diricinolein (1.0 mM) and ^{14}C-oleoyl-CoA (20 µM, 200,000 c.p.m.) and was incubated for 15 min at 30°C. The reactions were stopped and lipids were extracted using chloroform/methanol as previously described (5). Molecular species of acylglycerols were separated on a C18 column (25 × 0.46 cm, 5 µm, Ultrasphere C18, Beckman Instruments Inc., Fullerton, CA) using HPLC (8). Enzyme activity was determined based on the ^{14}C-label incorporated into the TAG products.

Lipase-Catalyzed Esterification

In order to evaluate which enzyme to use, reactions were carried out by mixing 40.8mg of glycerol adsorbed on silica gel (1:1, w/w) according to the approach of Berger et al. (9) and 0.2g of the 12-hydroxystearic acid (12-HSA) for a molar ratio of FFA to GL of 3:1. The reaction was performed at 85°C, just above the melting temperature of 12-HAS. At this temperature no organic solvent is necessary to solubilize the substrates, which allows for the use of a reaction medium solely composed of the necessary substrates. To optimize the reaction with the chosen enzyme, RMIM, several reaction variables were evaluated, and this paper presents the optimal conditions for synthesis of tri-12-hydroxystearoyl glycerol, trihydroxystearin (THS). The molar ratios of the FFA to adsorbed GL was 3:1, lipase loadingwas 10% (w/w) of the weight of FFA, $A_w$was 0.11, and the reaction temperature was 100°C. The reaction mixtures in an open glass vial (17mm i.d. × 85 mm L) was mixed vigorously by a magnetic stirrer (150 rpm) at the equilibrated temperature of 100°C using immobilized enzymes which were allowed to equilibrate at Aw of 0.11 prior to reaction. At various times during incubation, two 40µl samples were withdrawn from each flask and mixed with 0.4 ml of 2-propanol to dissolve fractions. The fractions were stored in a freezer at –20°C until HPLC analysis. Degree of esterification (%) represents the percentage of initial FA consumed in the reaction mixture.

HPLC Analysis

The course of esterification was monitored by reverse phase HPLC (*3*). Expected products of lipase reaction, i.e., 12-HSA, MG (monoacylglycerol), DG, and TG, were analyzed by injection of 50μL of sample on a RP-HPLC column, using mobile phase of methanol (A) and methanol water (90:10) (B). The gradient was as follows: from 100% B to 100% A in 20 min, then held for 12 min, and gradient back to 100% B in 2 min, then held for 6 min. The total run time was 40 min. The flow rate was 1.0 mL/min., and detection was performed at 205 nm. Results are expressed as percentage of peak areas. Retention times for MG, 12-HAS, DG, and TG were 4.94 min, 7.52 min, 20.46 min, and 32.34 min, respectively, and identity of each product confirmed by LC/MS.

Synthesis of 1,2 (2,3) Diricinoleoyl Glycerol

A general reaction path for lipase-catalyzed hydrolysis of triricinolein to diricinolein is shown in Figure 1.

Lipases usually hydrolyze the outer positions first (sn-1 or sn-3), as these are more available than the inner, sn-2 position. Once 2-MAG has been formed, most lipases easily break the last ester bond to form glycerol and FFA.

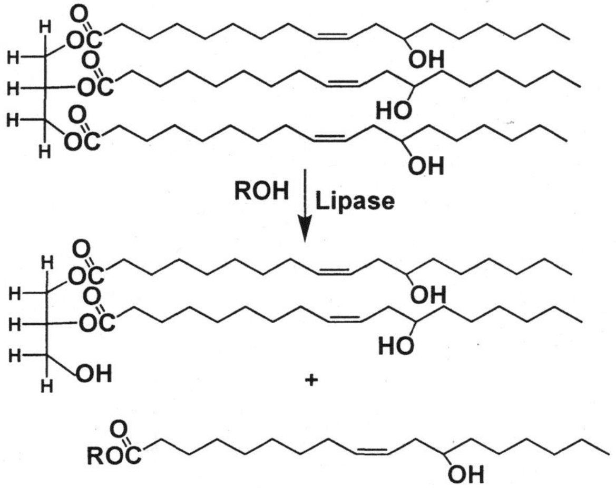

Figure 1. Reaction scheme for lipase-catalyzed hydrolysis/methanolysis of triricinolein to 1,2(2,3)-diricinolein and FFA/FAME.

Moreover, acyl migration can occur to form 1,3-DAG from 1,2(2,3)-DAG and 1(3)-MAG from 2-MAG. This type of migration is promoted in protic hydrophilic solvents like ethanol and methanol and aprotic hydrophobic solvents like hexane, but is slower in dipolar hydrophobic solvents like ethers and ketones (11). The purpose of this work was to find a lipase that selectively catalyzes this reaction, and does not further hydrolyze diricinolein to monoricinolein and glycerol. Therefore, the desired lipase should either show very low activity with 1,2(2,3)-diricinolein or give high yields of diricinolein before further hydrolysis is initiated. Additionally, the reaction was carried out in a solvent that would minimize acyl migration.

HPLC Analysis

Progress and yield of the reaction were monitored by determining the concentrations of residual triricinolein and ricinoleic acid, methyl ricinoleate and diricinolein formed. Monoricinolein was not found in the fractions, indicating that the fatty acid on the glycerol backbone was rapidly methanolyzed to give free glycerol and methyl ricinoleate. This effect has also been observed by others (10).

1,2- and 2,3-diricinolein are optical isomers (enantiomers), and are indistinguishable by ordinary HPLC techniques, although they can be converted to derivatives such as 3,5-dinitrophenyl urethanes, and separated using chiral HPLC (11). However, 1,3-diricinolein is a structural isomer which is readily separated from the 1,2(2,3)-isomers. Figure 2 shows the separation of 1,3- and 1,2(2,3)-diricinolein isomers. This system was also used for collection of small amounts of pure 1,2(2,3)-diricinolein.

Choice of Lipase

Four different lipases were tested in this study: CALB, RML, PCL and PRL. In a previous study on ethanolysis of vitamin A esters in DIPE and hexane, CALB, RML and PCL showed the best activity and stability of six lipases studied (12). PRL was also selected for the present study, since it has demonstrated high specificity for conversion of triacylglycerols to 1,2(2,3)-diacylglycerols (10).

Experiments with each of the four lipases were performed to evaluate effects of solvents (hexane and DIPE) as well as water activities (0.11, 0.53 and 0.97) on the yield of 1,2(2,3)-diricinolein (n=2). It was found that PRL had a higher specificity for the desired reaction than the other lipases investigated. This is demonstrated in Figure 3, and results are summarized in Table 2.

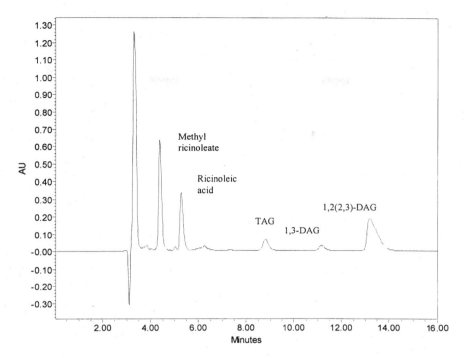

Figure 2. NP-HPLC analysis of a fraction from the PRL-catalyzed reaction in DIPE at a a$_w$ of 0.53. A CN column was used and a mobile phase consisting of hexane and MTBE (see the experimental section). The peaks were identified as methyl ricinoleate, ricinoleic acid, triricinolein (TAG), 1,3-diricinolein (1,3-DAG) and 1,2(2,3)-diricinolein (1,2(2,3)-DAG).

For CALB in DIPE at a$_w$ 0.11 when CALB, the highest yield of diricinolein, 40%, occurs after one hour of reaction, and then it is rapidly consumed for further production of methyl ricinoleate. RML and PCL were similar. This is usual for lipases, as they generally drive the reaction towards degradation of TAGs to either glycerol (non-specific lipases) or 2-MAGs (1,3-specific lipases) and FFAs or FAMEs. Figure 3, however, shows the yield of diricinolein from PRL catalyzing the reaction in the same solvent and the same A$_w$. In this case, the reaction is less driven towards formation of diricinolein. The yield reaches a maximum of 93% after around 24 hours of reaction, and is stable for another 24 hours. This result confirms that PRL selectively converts TAGs to DAGs but not further (13). The maximum yields of diricinolein for all the tested reaction conditions are given in Table 1. PRL is clearly the preferred lipase for the reaction, yielding between 63 and 93% for a$_w$ of 0.11 and 0.53 in hexane and DIPE.

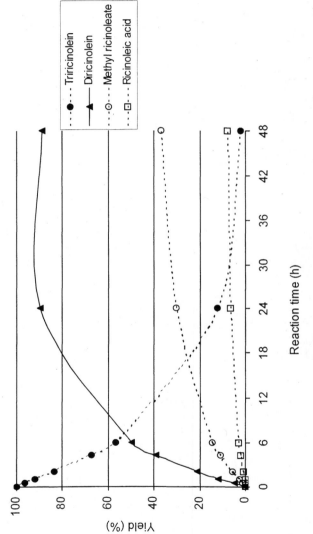

Figure 3. *Yield of diricinolein versus reaction time for methanolysis of triricinolein in DIPE at a_w 0.11, catalyzed by PRL (n=2).*

Table II. Recovery of Diricinolein. (%)

Immobilized Lipase	$a_w = 0.11$		$a_w = 0.53$		$a_w = 0.97$	
	Hexane	DIPE	Hexane	DIPE	Hexane	DIPE
CALB	40 (1h)	40 (1h)	16 (4h)	26 (6h)	20 (6h)	17 (4h)
RML	23 (1h)	21 (1h)	14 (1h)	37 (1h)	10 (48h)	36 (1h)
PCL	43 (7h)	33 (4h)	37 (1h)	36 (4h)	21 (1h)	40 (4h)
PRL	83 (48h)	93 (24h)	63 (48h)	88 (48h)	29 (48h)	49 (48h)

Recovery of diricinolein in (%) for each different reaction condition tested (n=2). The reaction time for maximum yield is shown in parentheses.

Hexane and DIPE were evaluated as reaction media for this study, since these solvents are commonly used for lipase-catalyzed reactions. In general, solvents of higher logP-value (such as hexane: logP~3.9) cause less inactivation of the enzyme than solvents of low logP-value (such as ethanol and acetonitrile: logP<0.2) (13). However, reaction rates are faster in solvents of lower logP-value. Hence, a solvent of intermediate logP-value, such as DIPE (logP~1.7), may be a good compromise. The DIPE provided slightly faster conversion of triricinolein to diricinolein using PRL (Table 1), while the maximum yield of diricinolein in hexane required (83%) 48 hours of reaction. Therefore, DIPE was considered the optimal solvent for the reaction catalyzed by PRL. An additional advantage of DIPE is reduced acyl migration vs. hexane.

Lipase activity at three different a_w were studied, 0.11, 0.53 and 0.97. The lower a_w gave a higher yield of diricinolein using any of the enzymes in any of the solvents. However, RML and PCL in DIPE, as shown in Table 1, produced diricinolein at higher yield with higher a_w. This is in agreement with work demonstrating that the initial reaction rate of PCL in DIPE was faster at higher a_w (17).

The PRL-catalyzed reaction in DIPE at a_w 0.53 and 0.97, is shown in Figure 4A and B, respectively. It can be seen that within 48 h, the higher yields of diricinolein were obtained at a_w 0.11 and 0.53 (93 and 88%, respectively). At the highest a_w (0.97 in Figure 4 B), the reaction was very slow, and maximum yield was not reached even after 48 hours of reaction. Since the reaction yields were similar at a_w 0.11 and 0.53, these two experiments were further evaluated. Although reaction at a_w 0.11 gave the higher yield of diricinolein, the proportion of 1,2(2,3)-isomer vs. 1,3-isomer was only 73%, compared to 88% for a_w 0.53. Thus, an a_w of 0.53 was selected as optimal for the reaction. By using NP-HPLC, the 1,2(2,3)-isomer is readily separated from the 1,3-isomer, thereby yielding >99% 1,2(2,3)-isomer (4).

In order to test the isolated 1,2(2,3)-diricinolein, it was used in an assay of *Arabidopsis thaliana* DGAT (AtDGAT). The DGAT catalyzes the final step in the Kennedy pathway for triacylglycerol biosynthesis, using acylCoA to acylate 1,2-diricinolein. The cloned AtDGAT cDNA was expressed in yeast *S.*

50

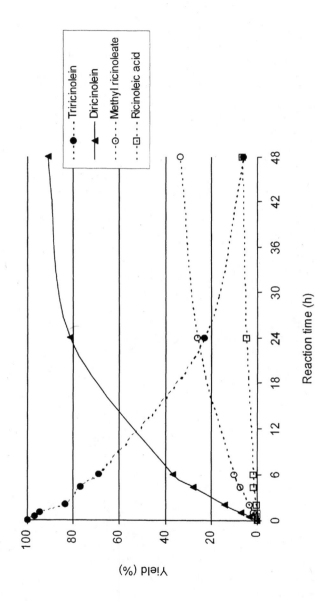

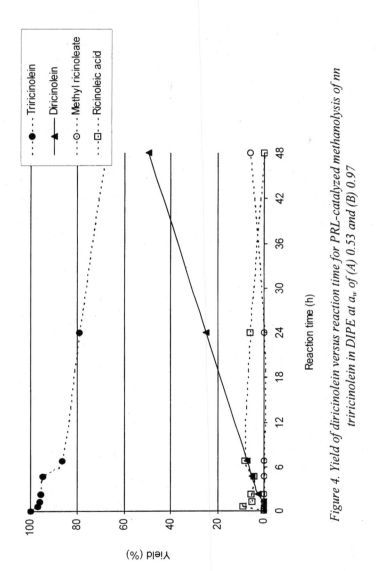

Figure 4. *Yield of diricinolein versus reaction time for PRL-catalyzed methanolysis of nn triricinolein in DIPE at a_w of (A) 0.53 and (B) 0.97*

cerevisiae, INVSc-1 strain using pYES2.1 TOPO TA Expression Kit. Microsomes containing AtDGAT proteins were extracted and used for DGAT activity assay. The results are shown in Figure 5.

The DGAT activity was significantly higher when we supplied the microsomes with 1 mM 1,2(2,3)-diricinolein (510 pmol/min/mg protein with diricinolein compared to 330 pmol/min/mg protein without added diricinolein). The incorporation of ^{14}C-labelled oleoyl-CoA into ^{14}C-labeled RRO increased from 0 in the absence of diricinolein to 290 pmol/min/mg protein in the presence of 1 mM diricinolein. The RRO was not produced in wild yeast cells when the same amount of 1,2(2,3)-diricinolein was supplied (data not shown). These results indicate that the diricinolein generated by lipase-catalyzed methanolysis is suitable as a substrate for the measuring DGAT activity.

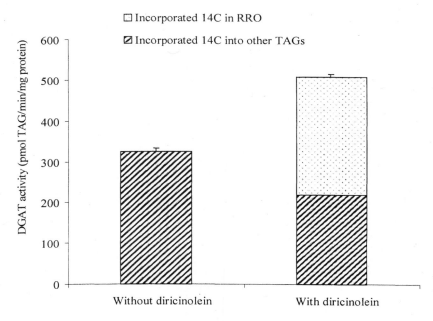

Figure 5. DGAT enzymatic activities in microsomes of yeast cells expressing AtDGAT. Data represent the mean of three experiments.

Synthesis of Tri 12-Hydroxystearoyl Acyl Glycerol (THS)

Figure 6 is a simple depiction of the desired reaction, using immobilized lipase to catalyze the condensation of three moles of 12-HSA to 1 mole of glycerol. Several immobilized lipases were evaluated to select the best choice.

The RMIM, a 1,3-specific lipase, was selected as the lipase for the present study based on the yield of 71% TG, highest obtained from the seven lipases studied (Table III). Although Novozym 435 (immobilized CALB) showed the highest degree of esterification, the yield of TG was a less than that for RMIM. Although the lipases tested are all 1,3-specific, it has been suggested (*14*) that a 1,3-specific lipase would catalyze the synthesis of TG via acyl migration. Thus, the synthesis of TG should proceed most readily under conditions that promote acyl migration, such as high temperature and in a protic medium.

Conditions for optimum esterification were determined for RMIM, with a molar ratio of 12-HAS to glycerol of 3:1 and A_w 0.11 reacted at 100°C for 24 h with stirring. Since reactions were conducted at 100°C, water evolved at

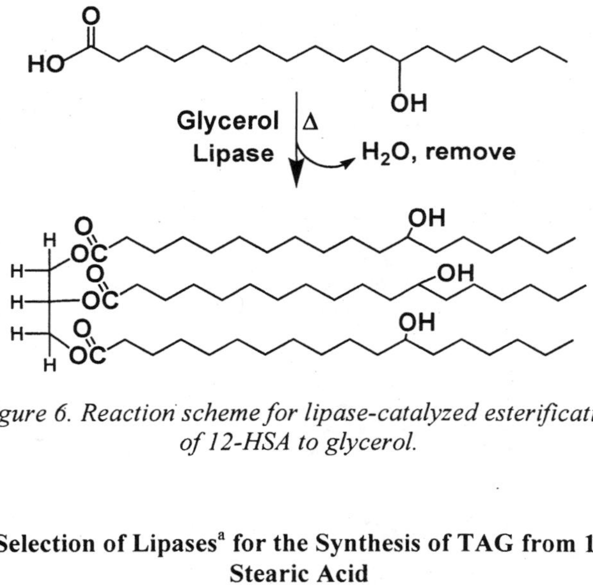

Figure 6. Reaction scheme for lipase-catalyzed esterification of 12-HSA to glycerol.

Table III. Selection of Lipases[a] for the Synthesis of TAG from 12-Hydroxy Stearic Acid

Enzymes	MG	12-HSA	DG	TG
RMIM	1.1	9.0	18.9	71.0
Novozym 435	2.6	7.0	25.4	67.6
CRL	4.8	91.7	2.8	0.6
PRL	11.6	81.5	5.3	1.6
ROL	12.2	78.2	8.1	1.5
ANL	6.4	90.3	3.3	0.0
PCL	3.4	95.5	1.1	0.0

RMIM, Novozym 435 and all immobilized on a polypropylene powder were stored at an A_w of 0.11 and reacted at 85°C for 24 hr as described in the Experimental section.

54

ambient pressure during condensation reactions. Detailed changes in glyceride and FFA composition of the reaction mixture during the course of esterification were monitored by HPLC and are shown in Figure 7. Initially, the rate of DG synthesis was high with 56% accumulating, but after 1hr the concentration of DG decreased. There was a decrease in the content of FA to 9% during the first 1hr, indicating the initial esterification occurs rapidly. The concentration of MG remained low (less than 2 %) throughout the reaction. Apparent equilibrium was reached within 12 h with a TG concentration of 76%. Over 98% of FA was converted to glycerides.

Enzymatic production of tri-12-hydroxystearin serves as an example of an alternative to chemical reaction for producing triacylglycerols with nonfood applications, such as lubricants, coatings, and cosmetics. To date, most structured lipids have been synthesized for food or nutraceutical applications, but there is also potential for this technology in non-food applications. Conversion of THS to the diacylglycerol is currently underway.

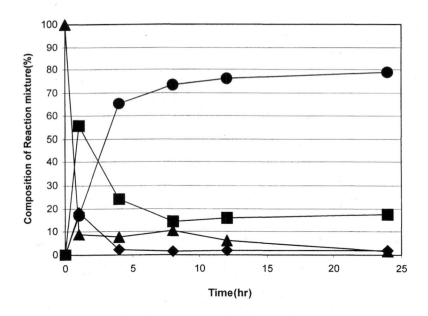

Figure 7. The composition of the reaction mixture during enzymatic esterification of 12-hydroxystearic acid and glycerol at Aw=0.11; Reaction conditions were as follows: 40.8 mg of GL adsorbed on silica gel (weight ratio of glycerol to adsorbent =1:1), 0.2g of 12-hydroxystearic acid, molar ratio of FFA to glycerol of 3:1, 20mg of RMIM (A_w 0.11), stirred at 150rpm and 100 (C. Monohydroxystearin (MG, ◆); 12-Hydroxystearic acid (FFA, ▲); Dihydroxystearin (DG, ■); and Trihydroxystearin (TG, ●).

Summary

1,2-Diacylglycerols (DG) are the native substrates for the diacylglycerol acyltransferase (DGAT). While DG containing saturated or unsaturated fatty acids are relatively easy to make chemically, it is difficult to chemically synthesize DG containing hydroxy fatty acids in specific positions on the glycerol backbone. An alternate approach is to start from acylglycerols containing hydroxy fatty acids, and selectively remove fatty acid chains to obtain the desired acylglycerol. Our goal was to produce both natural and "unnatural" substrates to evaluate the substrate specificity of the DGAT. This study identified optimal parameters for lipase-catalyzed methanolysis of triricinolein to produce 1,2(2,3)-diricinolein, the substrate for biosynthesis of triricinolein in castor (*Ricinus communis*). We tested four different immobilized lipases, using n-hexane and diisopropyl ether (DIPE) as reaction media, and three different water activities. We followed the consumption of triricinolein and the formation of diricinolein, methyl ricinoleate and ricinoleic acid during the course of the reaction. *Penicillium roquefortii* lipase gave the highest yield of 1,2(2,3)-diricinolein and this lipase showed the highest specificity for the studied reaction, i.e. high selectivity for reaction with triricinolein, but low for diricinolein. The diacylglycerol produced can be acylated by the diacylglycerol acyltransferase from *Arabidopsis thaliana*. Therefore, the product of the lipase reaction is a suitable substrate for acyltransferase reactions. In order to produce an "unnatural" substrate, we evaluated conditions for synthesizing tri-hydroxy-stearin, using esterification of glycerol and 12-hydroxystearic acids in a solvent free system. We identified optimal parameters for lipase-catalyzed synthesis of tri-12-hydroxystearin (THS) from glycerol and 12-hydroxystearic acid (12-HSA), using Lipozyme RMIM 60 from *Rhizomucor miehei* (RML), a 1,3-specific lipase, as the biocatalyst. in this study. We obtained yields of THS up to 75% under conditions in which water was removed during the course of the reaction. Our future work with this product will be to produce 1,2 -di 12 hydroxy stearoyl glycerol.

Acknowledgments

We would like to thank USDA-IFAFS for Grant no. 2000-04820, which provided support for Charlotta Turner. We also gratefully acknowledge support for Xiaohua He and Sung-Tae Kang from Dow Chemical Company through CRADA 58-3K95-2-918. We are grateful to Novozymes A/S for generously providing the Lipozyme RM-IM and Novozyme 435 enzymes, and Amano Enzymes Inc. for providing Lipase PS and Lipase RG.

References

1. Caupin, H.-J. In *Lipid Technologies and Applications;* Gunstone, F. D.; Padley, F. B., Eds.; Marcel Dekker: New York, 1997; pp. 787-795.

2. McKeon, T. A. In *Industrial Uses of Vegetable Oil*; Erhan, S. Z.; Ed.; AOCS Press, Champaign, IL, 2005; pp. 1-13.

3. He X.; Turner C.; Chen G. Q.; Lin J. T.; McKeon T. A. *Lipids,* **2004**, *39*, 311-318.

4. Turner C.; He X.; Nguyen T.; Lin J. T.; Wong R. Y.; Lundin, R. E.; Harden L.; McKeon T. *Lipids,* **2003**, *38*, 1197-1206.

5. Klibanov, A. M. *Acc. Chem. Res.* **1990**, *23*, 114-120.

6. Halling, P. J. *Biotechnol. Tech.* **1992**, *6*, 271-276.

7. McKeon, T. A.; Lin, J. T.; Goodrich-Tanrikulu, M.; Stafford, A. E. *Ind. Crops Prod.* **1997**, *6*, 383-389.

8. Lin, J.-T.; Woodruff, C. L.; McKeon, T. A. *J. Chromatogr. A.* **1997**, *782*, 41-48.

9. Berger, M., Laumen, K.; Schneider, M. P. *J. Am. Oil Chem. Soc.* **1992**, *69*, 955-960.

10. Fureby, A. M.; Tian, L.; Adlercreutz, P.; Mattiasson, B. *Enzyme Microb. Technol.* **1997**, *20*, 198-206.

11. Itabashi, Y.; Kukis, A.; Marai, L.; Takagi, T. *J. Lipid Res.* **1990**, *31*, 1711-1717.

12. Turner, C.; Persson, M.; Mathiasson, L.; Adlercreutz, P.; King, J.W. *Enzyme Microb. Technol.* **2001**, *29*, 111-121.

13. Adlercreutz, P. 2000. In *Applied Biocatalysis* Straathof, A. J. J.; Adlercreutz, P., Eds.; Harwood Academic Publishers, Newark, NJ pp. 295-316.

14. Lortie, R.; Trani, M.; Ergan, F. *Biotechnol. Bioengin.* **1993**, *41*, 1021-1026.

Chapter 4

Controlled Release of Indomethacin from Spray-Dried Chitosan Microspheres Containing Microemulsion

Jun Bae Lee[1], Hyun Jung Chung[1], Somlak Kongmuang[2], and Tae Gwan Park[1,*]

[1]Department of Biological Sciences, Korea Advanced Institute of Science and Technology, Daejeon 305-701, South Korea
[2]Department of Pharmaceutical Technology, Silpakorn University, Nakornpathom, Thailand, 73000

Indomethacin microemulsion loaded chitosan microspheres (IMCMs) were prepared by a spray-drying technique. IMCMs were spheroids formed by aggregates of microspheres with sizes of ~8 μm, while indomethacin loaded chitosan microspheres (ICM) (control) did not show any agglomeration. Indomethacin (IDM) appeared to be in an amorphous state in both ICMs and IMCMs, confirmed by X-ray diffraction and differential scanning calorimetry. The encapsulation efficiency of IDM was higher in IMCM than ICM. IMCMs exhibited a pH independent and sustained release pattern with small burst effect, compared to ICMs. Thus, drug-in-microemulsion loaded polymeric microspheres could possibly provide a sustained release formulation.

Various drugs have been encapsulated within polymeric microspheres by different techniques for drug delivery applications. One particular interest is an oral sustained released formulation. Recently, nanotechnology has attracted considerable attention by introducing nanoscale drug carriers for targeted drug delivery or sustained release. Microemulsion technique is one method for obtaining particles of nano-size. A microemulsion is a thermodynamically stable, isotropically clear dispersion of two immiscible liquids stabilized by an interfacial film of surfactant molecules (1). Microemulsions are mainly used to solubilize a poorly water soluble compound in water by the addition of an emulsifier and co-emulsifier(s). Thus, by using this technique, bioavailabilities of poorly absorbed drugs were dramatically enhanced through increasing their solubility (1-3). Interestingly, microemulsions have not only been used directly as drug delivery vehicles, but have also been used indirectly as a means of producing solid drug-loaded nanoparticles by spraying of an o/w microemulsion (1, 4).

Chitosan is a naturally occurring polymer mainly found in marine crustacean, which is cationic, biodegradable and biocompatible, making it useful for many biomedical applications (5-10). In this study, a spray drying technique was applied for obtaining solid microparticles containing indomethacin. Spray drying is a well-known process used to obtain dry powders or agglomerates from solution or suspension systems of interest (11). One advantage of spray drying is that it produces particles with sizes in the range of a few microns to several tens of microns, having a narrow size distribution. Cimetidine and famotidine loaded chitosan microspheres produced by a spray drying method gave a fast release profile with a burst effect (11). Chitosan microspheres have also been applied as a mucoadhesive agent in formulations for sustained release systems (13-15). The objective of the present study was to prepare chitosan microspheres encapsulating indomethacin loaded microemulsion by the spray-drying technique for better controlling release patterns. Indomethacin was used as a model drug because of its low aqueous solubility. The loading efficiency of indomethacin (IDM) with or without the microemulsion in the chitosan microsphere was evaluated by comparing the two types of chitosan with different molecular weights. We also investigated the physicochemical properties of chitosan microspheres by varying formulation parameters and studied IDM release characteristics in vitro.

Materials and Methods

Materials

Chitosan (CS 100, $Mn = 2.7 \times 10^5$ and CS 500: $Mn = 7.16 \times 10^5$) and indomethacin was obtained from Wako (Tokyo, Japan) and Sigma (St. Louis, MO), respectively. Plurol Oleique cc 497 (Oil) and Labrasol (surfactant) were purchased from Gaffefossé (Saint-Priest, France). Ethanol, acetic acid and all other chemicals were obtained commercially and of analytical grade.

Preparation of IDM loaded chitosan microspheres (ICM) by the spray-drying technique

Chitosan was dissolved in 1 % (v/v) of aqueous acetic acid solution to obtain 0.25 % or 1 % (w/v) solution. Various amounts of IDM were dissolved in 15 ml of ethanol before mixing with the above solution. Each mixture was magnetically stirred for 4 h. The dispersed solution was fed with air and passed separately to the nozzle of a mini spray-dryer (Büchi B-191, Switzerland). The air caused the feed mixture to break up into droplets at the outlet of the nozzle. After evaporation of the solvent, chitosan microspheres were formed and collected with a cyclone separator. The inlet temperature was 140 °C with a mixed solution feed rate of 6 ml/min. The air flow rate was 600 l/h with aspiration of 90%. The outlet temperature was maintained between 95 and 105 °C. The spray-dried products were retrieved from the collecting chamber and stored in a desiccator at room temperature.

Preparation of IDM loaded chitosan microemulsion system (IMCM)

One hundred milligrams of IDM was dissolved in 1 g of Plurol Oleique cc 497 as the oily-reservoir. The oil phase was added into 100 ml of chitosan-acetic acid solution (1 % chitosan concentration) containing 6 g of Labrasol as a surfactant. The microemulsion solution was placed under magnetic stirring condition for 4h. The clear solution of the mixture was analyzed by a laser light scattering technique (Zetaplus, Brookhaven Instrument Corp., USA). This homogeneous microemulsion solution was spray-dried.

Scanning electron microscopy (SEM)

Scanning electron microscopy (SEM, Philips 535M) was used for observing the morphology of all chitosan microspheres. The samples were coated with gold using a sputter-coater (HUMMER V, Technics, USA). Argon gas pressure was set at 5 psig with the current maintained at 10 mA for 5 min.

Drug content determination

The loading percentage of IDM within the chitosan microspheres with and without the oily-reservoir was determined after dissolving 20 mg of microspheres in 20 ml of chloroform and methanol, respectively. The solution was sonicated for 30 min to extract the inner drugs from the microspheres. The drug content was analyzed with the spectrophotometer UV 1601 (Shimadzu, Japan) at a wavelength of 320 nm using solvent as a blank.

Differential scanning calorimetry (DSC)

DSC was measured using a Du-Pont 9900 instrument (Model 2910). DSC was performed for pure IDM, chitosan and ICM with or without an oily reservoir. Samples were purged with an atmosphere of nitrogen. The heat flow rate was recorded at temperatures from 100 °C to 200 °C at a rate of 10 °C min^{-1}.

X-ray diffraction (WAXD) study

The crystallinity of chitosan particles, pure IDM, and ICM with or without an oily reservoir was examined by X-ray analysis. WAXD was investigated using a wide angle X-ray diffractometer (Rigaku, Japan) at a scanning range of 2θ from 3 to 60 ° at ambient temperature using Cu-Kα radiation. The scanning rate was 3 ° (2θ)/min.

In vitro release studies

In vitro release tests were carried out by following the applied USP paddle method at a stirring speed of 100 rpm in a beaker. Twenty mg microspheres with an oily-reservoir were immersed in 900 ml of pH 7.4 phosphate buffered saline solution (PBS) at 37 °C. Ten ml of sample was withdrawn at different time intervals and replenished immediately with the same amount of fresh buffer solution. The withdrawn samples were separated by ultracentrifugation (Beckman, USA) at 12,000 rpm for 20 min and the supernatants were analyzed using the UV-Visible spectrophotometer UV 1601 (Shimadzu, Japan) by detection at 320 nm. ICMs without an oily reservoir were used as a control. Furthermore, the effect of pH on the release profiles was investigated by immersing the chitosan microspheres in a pH 1.5 HCl/KCl solution following the above same paddle method. All release tests were run in triplicate and the mean value were calculated.

Results and Discussion

Morphology of microspheres

IMCMs and ICMs were obtained by the spray-drying method as shown in Figure 1. SEM analysis revealed that all micropsheres were spherical (Figure 2).

ICM appeared as well separated particles with a diameter of 3~4 µm. Its surface indentations could be attributed to subsequent shrinking of the microsphere after the solid crust was formed (16). The IMCMs appeared as aggregates of spherical particles with a sizes of 6~8 µm. The different surface characteristics of IMCM could result from the incorporation of the microemulsion in the formulation. These IMCMs could flow but somewhat adhered to the collecting chamber, decreasing the yield of IMCMs. The single phase system (ICM) produced smaller and individual microspheres than the two phase system (IMCM). Similar observations were also reported (12, 17, 18).

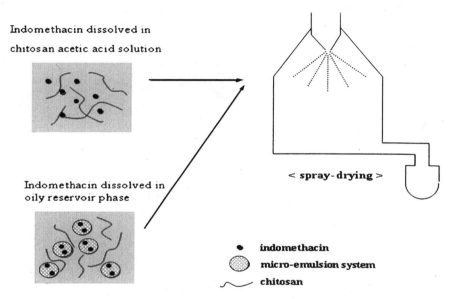

Indomethacin dissolved in
chitosan acetic acid solution

Indomethacin dissolved in
oily reservoir phase

< spray- drying >

● indomethacin
micro-emulsion system
chitosan

Figure 1. Schematic illustration of the spray-drying process

Drug Loading and Encapsulation Efficiency

ICMs and IMCMs were prepared using the same amount of IDM in chitosan at 10% (w/w). Drug loading amount was 7.0 ± 0.3 % (w/w) for ICM, while that for IMCM was 8.0 ± 0.2 % (w/w). Also, the encapsulation efficiency increased from 76.5 ± 3.9 for ICM, to 87.4 ± 2.9 for IMCM. The higher values for IMCM could be attributed to the use of a microemulsion which could have increased the solubility of IDM by acting as an oil reservoir. IDM could possibly partition to a more extent within the microemulsion droplet phase. This could have caused IDM to be more soluble in the microemulsion for IMCM than in the aqueous chitosan solution for ICM.

Figure 2. Scanning electron micrographs of (a) ICMs and (b) IMCMs prepared with 1 % (w/v) CS 100 at 10 % (w/w) drug/polymer.

Characterization of IDM in Microspheres

The WAXD spectra for IDM crystal (a), ICM (10% w/w) (b), IMCM (10% w/w) (c) and chitosan powder (d) are presented in Figure 3 (*1*). The spectra for IDM include an intense peak between 2θ of 15° and 30° due to its crystallinity, while that of the chitosan powder shows an amorphous state. These results show evidence that IDM exists in an amorphous state in both microspheres and oily reservoir microspheres. The DSC tracings for pure IDM (a), ICM (b), IMCM (c) and pure chitosan are presented in Figure 3 (*2*). The endothermic peak of pure IDM was observed at 161.8 °C which represented the most stable IDM γ-form (*20*). No peak appeared in the DSC thermograms of (b), (c) and (d) since IDM could be incorporated within chitosan at a molecularly dissolved state as an evidence of WXAD. We did not observe glass transition temperature (Tg) of chitosan in the temperature range of 120 -190 °C.

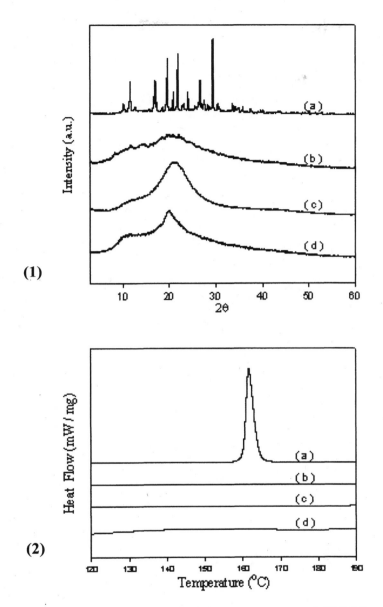

Figure 3. (1) Wide angle X-ray diffraction patterns and (2) DSC analysis of (a) IDM powder, (b) ICM, and (c) IMCM (10 % (w/w) drug/polymer), and (d) chitosan powder (CS 100).

In vitro Drug Release

In vitro drug release was performed in either pH 7.4 PBS solution or pH 1.5 HCl/KCl solutions for 8 hours. Results for percent of drug release in pH 7.4 PBS versus time for ICM and IMCM (10% w/w drug/polymer) are presented in Figure 4. ICMs showed rapid release of IDM with a burst release within 2 h. Thus, we suspected that IDM might possibly be located at the surface of microsphere. Also, the IDM dissolution profiles reached a plateau, about 90% release within 4 h. These results can be explained by the property of IDM in the amorphous state (*21, 22*), which happened to exist in its most unstable and soluble form. We proposed that the potential release mechanism of IDM from ICM in pH 7.4 PBS was a matrix-type diffusion mechanism including three steps: (i) water uptake and swelling of microsphere, (ii) dissolution of IDM and (iii) diffusion of IDM through the chitosan microsphere gel (*6*).

However, the dissolution profile for IMCM in pH 7.4 PBS demonstrated a significantly lower IDM release rate compared to ICM. We did notice that there was a small burst release, but otherwise showed a much more sustained release profile. The swollen chitosan gel layer, even though releasing out a small burst

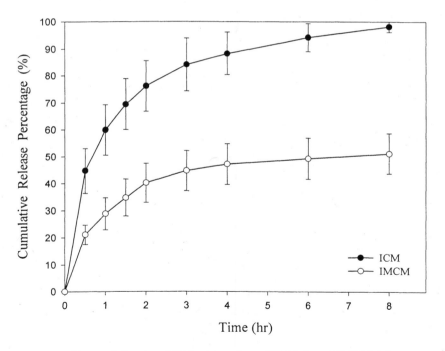

Figure 4. In vitro IDM release from ICM (●) and IMCM (○) in pH 7.4 PBS solution.

of IDM initially, could retard IDM dissolution which resulted in the sustained release pattern. The slow IDM release could also be due to IDM partitioning and diffusion from the oily reservoir through the chitosan gel.

A burst release was also observed for ICM in pH 1.5 medium, as shown in Figure 5. Immersing the ICMs in acidic medium would result in swelling and dissolution of chitosan, causing most of the IDM to rapidly dissolve out (5). In this case, IDM release from the chitosan matrix might involve three mechanisms: (i) IDM release from the surface of the particle, (ii) IDM diffusion through the swollen rubbery matrix and (iii) IDM release due to polymer erosion (5).

IMCM in pH 1.5 medium showed similar drug release profiles as in pH 7.4. It seemed that the microemulsion within IMCM was effective in sustained release, especially at low pH conditions. Thus, the rate limiting step of IDM dissolution would have been from IDM partitioning through the microemulsion. IDM seemed to have dissolved in the hydrophobic part of the microemulsion and oil reservoir which, on contact to the medium, partitioned into the medium. Thus, the microemulsion system brought out sustained release profiles of IDM, independent of pH.

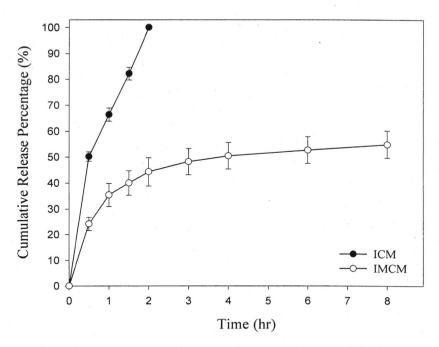

Figure 5. In vitro IDM release from ICM (●) and IMCM (○)
in pH 1.5 HCl/KCl solution.

Conclusion

This study demonstrated that IMCM could be easily prepared by a spray drying method. The amount of IDM loading and encapsulation efficiency was higher for IMCM than ICM, since the microemulsion system increased the solubility of IDM by acting as a surfactant. The dissolution profiles of IDM from the ICM formulation in pH 7.4 PBS did not show sustained release patterns, and a "burst release" was observed in acidic medium. These dissolution profiles could be an effect caused by the IDM amorphous state and chitosan solubility. The amorphous state of IDM within ICM and IMCM was confirmed by WAXD and DSC. In the case of IMCM, IDM seemed to partition from the oily reservoir resulting in sustained release patterns in both mild and acidic pH conditions. Since the application of microemulsions in oral delivery should concern toxicity of the various surfactants, further investigations *in vivo* should be performed.

Acknowledgements

We are thankful to the nano-biomaterials laboratory, KAIST for providing facilities. We would also like to thank the Royal Thai Government for providing a grant for Dr. Somlak Kongmuang to be at KAIST.

References

1. Lawrence, M. J.; Rees, G. D. *Adv. Drug. Del. Rev.* **2000**, *45*, 89.
2. Kim, C. K.; Cho, Y. J.; Gao, Z. G. *J. Control Release* **2001**, *70*, 149.
3. Sha, X.; Yan, G.; Wu, Y.; Li , J.; Fang, X. *Eur. J. Pharm. Sci.* **2005**, *24*, 477.
4. Andersson, M.; Lofroth, J. E. *Int. J. Pharm.* **2003**, *257*, 305.
5. Agnihotri, S. S.; Amninabhavi, T. M. *J. Control Release* **2004**, *96*, 245.
6. Asada, M.; Takahashi, H.; Okamoto, H.; Tanino, H.; Danjo, K. *Int. J. Pharm.* **2004**, *270*, 167.
7. Ribeiro, A. J.; Neufeld, R. J.; Arnaud, P.; Chaumeil, J. C. *Int. J. Pharm.* **1999**, *187*, 115.
8. Lorenzo-lamosa, M. L.; Remunn-Lopez, C.; Villa-Jato, J. L.; Alonso, M. J. *J. Control Release* **1998**, *52*, 108.
9. Munjeri, O.; Collett, J. H.; Fell, J. T. *J. Control. Release.* **1997**, *46*, 273.
10. Giunchedi, P.; Juliano, C.; Gavini, E.; Cossu, M.; Sorrenti, M. *Eur. J. Pharm. Biopharm.* **2002**, *53*, 233.
11. He, P.; Davis, S. S.; Illum, L. *Int. J. Pharm.* **1999**, *187*, 53.
12. Martinac, A.; Filipovic-Grcic, J.; Voinovich, D.; Perissuti, B.; Franceschinis, E. *Int. J. Pharm.* **2005**, *291*, 69.

13. Martinac, A.; Filipovic-Grcic, J.; Babaric, M.; Zorc, B.; Voinovich, D.; Jalsenjak, I. *Eur. J. Pharm. Sci.* **2002**, *17*, 207.

14. Ko, J. A.; Park, H. J.; Hwang, S. J.; Park, J. B.; Lee, J. S. *Int. J. Pharm.* **2002**, *249*, 165.

15. Berthod, A.; Cremer, K.; Kreuter, J. *J. Control. Release.* **1996**, *39*, 17.

16. Rege, P. R.; Garmise, R. J.; Block, L. H. *Int. J. Pharm.* **2003**, *252*, 41.

17. Carli, F.; Chiellini, E. E.; Bellich, B.; Macchiavelli, S.; Cadelli, G. *Int. J. Pharm.* **2005**, *291*, 113.

18. He, P.; Davis, S. S.; Illum, L. *Int. J. Pharm.* **1998**, *166*, 75.

19. Xu, Y.; Du, Y. *Int. J. Pharm.* **2003**, *250*, 215.

20. Slavin, P. A.; Sheen, D. B.; Shepherd, E. E. A.; Sherwood, J. N.; Feeder, N.; Docherty, R.; Milojevic, S. *J. Crytal. Growth.* **2002**. *237-239*, 300-305.

21. Giunchedi, P.; Genta, I.; Conti, B.; Muzzarelli, R. A. A.; Conte, U. *Biomaterials*, **1998**, *19*, 157.

22. Pohlmann, A. R.; Weiss, V.; Mertins, O.; Silveira, N. P.; Guterres, S. S. *Eur. J. Pharm. Sci.* **2002**, *16*, 305.

Chapter 5

Synbiotic Matrices Derived from Plant Oligosaccharides and Polysaccharides

Arland T. Hotchkiss, Jr.[1,*], LinShu Liu[1], Jeff Call[1], Peter Cooke[1], John B. Luchansky[1], and Robert A. Rastall[2]

[1]Eastern Regional Research Center, Agricultural Research Service, U.S. Department of Agriculture, 600 East Mermaid Lane, Wyndmoor, PA 19038
[2]School of Food Biosciences, University of Reading, P.O. Box 226, Whiteknights, Reading RG6 6AP, United Kingdom

A porous synbiotic matrix was prepared by lyophilization of alginate and pectin or fructan oligosaccharides and polysaccharides cross-linked with calcium. These synbiotic matrices were excellent structures to support the growth of *Lactobacillus acidophilus* and *Lactobacillus reuteri* under anaerobic conditions. When the matrix was inoculated with lactobacilli and stored at 4°C under aerobic conditions for a month, bacterial viability was preserved even though the synbiotic dried to a hard pellet.

Functional foods that benefit health by including live friendly bacteria as ingredients have a long history of use in Asia and Europe, and are rapidly emerging in the US. Synbiotics are a combination of probiotic, or health-promoting bacteria, and a prebiotic. Probiotic bacteria, such as *Bifidobacteria* and *Lactobacillus*, are facultative anaerobes found in the gut that promote host health. Prebiotics are components of non-digestible fiber that selectively stimulate the growth and/or activity of one or a limited number of these probiotic bacteria (*1*). Native probiotic bacteria can become depleted due to antibiotic therapy or diets low in soluble fiber and rich in refined carbohydrates (*2*). Dietary manipulation of the colonic microflora using these agents can restore the probiotic microflora, prevent pathogen colonization through competitive exclusion, stimulate the immune system and may play a role in the prevention of gastrointestinal diseases. Unfortunately, long term colonization of the gut by orally administered probiotic bacteria has been difficult to demonstrate. Additionally, viability of probiotic bacteria in commercial products varies. Therefore, prebiotics that selectively stimulate the growth of native probiotic bacteria is an attractive strategy to improve gut microflora composition and consequently host health. Another alternative is to supply synbiotics such that the probiotic growth is enhanced as much as possible in the colon. Supplying lactobacilli in the presence of fructo-oligosaccharides (FOS) produced higher daily weight gain and better feed conversion ratio for swine compared to feeds supplemented with the probiotic alone (3).

Pectic oligosaccharides (POS) are known to have prebiotic properties (4, 5). Since guluronic acid is an epimer of galacturonic acid, and both alginate and pectin can be crosslinked with calcium, we examined alginic acid-calcium and POS as a synbiotic matrix. Modified citrus pectin (MCP) is a low molecular weight, low-methoxy pectin rich in oligogalacturonic acids (6). However, it is not known if MCP has prebiotic properties. FOS and inulin were included in alginic acid-calcium matrices as examples of currently used commercial prebiotics.

Materials and Methods

A solution of high viscosity alginic acid (Type IV, Sigma; 10 mg/ml) and another oligo-/poly-saccharide (10 mg/ml) was prepared in deionized water. Other saccharides included orange peel pectic oligosaccharides (5), modified citrus pectin (Pectasol, EcoNugenics, Santa Rosa, CA), fructo-oligosaccharides (Raftilose P95, Orafti, Malvern, PA) and inulin (Raftilose Synergy1, Orafti, Malvern, PA). The solution was pipetted into a 96 well plate (120 μL/well) and then lyophilized. The dry matrix was immersed in a 0.5% $CaCl_2$ solution for 20 minutes. The matrix was then washed with deionized water (3x, stirring). The calcium cross-linked matrix was then returned to the 96 well plate and lyophilized. Finally, the matrix was washed with ethanol.

Cultures of *Lactobacillus acidophilus* (1426) and *Lactobacillus reuteri* (1428) were grown in deMan, Rogosa and Sharpe (MRS) broth (pH 5.5-6; Difco) under anaerobic conditions (5% H_2, 10% CO_2, 85% N_2, 37°C) to 10^9 cfu/mL. The cultures were diluted 10x and the matrix plugs were inoculated with 20 μL of 10^8 cfu/mL under aerobic conditions in the 96 well plate. Each week an inoculated matrix plug was placed in a culture tube containing MRS or Brain Heart Infusion (BHI) broth (pH 7; Difco) and returned to anaerobic conditions to monitor bacterial growth (visual turbidity determined by a trained laboratory technician). This analysis was performed in duplicate. A matrix control consisted of inoculating a culture tube with 20 μL of the 10^8 cfu/mL lactobacillus cultures. Another control was used in which an uninoculated matrix plug was added to a BHI culture tube. Inoculated matrix plugs were stored in the 96 well plate wrapped in parafilm at 4°C for up to a month under aerobic conditions.

Samples were processed for scanning electron microscopy (SEM) by fixing the matrix plugs with 2.5% glutaraldehyde in 0.1M imidazole buffer solution (pH 7.2) for 5 hours, dehydrating in an ethanol series (50%, 80%, 100%), freezing in liquid nitrogen, fracturing with a cold surgical scalpel blade, thawing fractured pieces into absolute ethanol (5-10 minutes), critical point drying from CO_2, and sputter coating with a thin layer of gold. Samples were examined with a Quanta 200 FEG environmental scanning microscope (FEI Co., Inc., Hillsboro, OR) operated in the high vacuum/secondary electron imaging mode.

Results and Discussion

Anaerobic bacterial growth of *L. acidophilus* and *L. reuteri* was observed for all of the inoculated matrices stored for up to a month at 4°C under aerobic conditions (Table I). In all cases there was bacterial growth in the control tubes inoculated with the lactobacilli without matrix and no bacterial growth was observed in BHI control tubes containing uninoculated matrix. Inoculated matrix stored for up to a month at 4°C under aerobic conditions had shrunk to a hard pellet. Yet the matrix was able to preserve the viability of the lactobacilli for a month under these conditions.

Table I. Growth of Lactobacilli in Carbohydrate Matrices

Time	Alg-Ca		POS-Alg-Ca		MCP-Alg-Ca		FOS-Alg-Ca		Syn-Alg-Ca	
(days)	1426	1428	1426	1428	1426	1428	1426	1428	1426	1428
0	+,+	+,+	+,+	+,+	+,+	+,+	+,+	+,+	+,+	+,+
7	+,+	+,+	+,+	+,+	+,+	+,+	+,+	+,+	+,+	+,+
14	+,+	+,+	+,+	+,+	+,+	+,-	+,+	+,+	+,+	+,+
21	+,+	+,+	+,+	+,+	+,+	+,+	+,+	+,+	+,+	+,+
30	+,+	+,+	+,+	+,+	+,+	+,+	+,+	+,+	+,+	+,+

Alg-Ca = Alginic acid-calcium

POS = Pectic oligosaccharides

MCP = Modified citrus pectin

FOS = Fructo-oligosaccharides

Syn = Inulin

1426 = *Lactobacillus acidophilus*

1428 = *Lactobacillus reuteri*

+ = growth

- = no growth

Scanning electron microscopy of the alginic acid-calcium matrix revealed a honeycomb-like structure (Figure 1) very similar to the pectin-calcium matrix structure reported previously (7). This structure included many cavities for growth of lactobacilli (Figure 2). Since alginic acid is not known to have prebiotic properties, the MRS culture medium was responsible for this growth of lactobacilli. When pectic oligosaccharides and modified citrus pectin were included in the alginic acid-calcium matrix, the honeycomb-like structure was more compact with smaller internal cavities (Figure 3). The galacturonic acid content of MCP was higher than that present in POS (5, 6), making MCP potentially more crosslinked by calcium. The surface of the MCP-alginic acid-calcium matrix was more rough and bumpy (Figure 4), due to the calcium cross-linking of both pectin and alginate networks. Lactobacilli were only observed scattered on the surface of the POS- and MCP-alginic acid-calcium matrix (Figure 4) after three days of growth in BHI broth with inoculated matrix plugs stored at 4°C under aerobic conditions for two weeks. The BHI broth does not support as active growth of lactobacilli compared to MRS which is selective for lactobacilli. When POS- and MCP-alginic acid-calcium matrix plugs, stored at 4°C under aerobic conditions for three weeks or more, were placed in MRS broth under anaerobic conditions, lactobacilli grew so rapidly that the plugs disintegrated in a day of growth. Therefore, these samples were not processed for electron microscopy. When FOS and inulin were used in the alginic acid-calcium matrix, the internal cavities were packed with lactobacilli (Figure 5) after a day of growth in MRS broth. Bacterial growth expanded in the FOS- and inulin-alginic acid-calcium matrices such that the matrix structure began to break (Figure 6) and the plugs eventually fell apart.

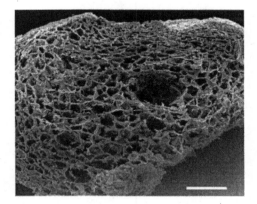

*Figure 1. General SEM image of the alginic acid-calcium cross-linked matrix plug four days after transfer of the inoculated (*Lactobacillus acidophilus*) plug to BHI broth and anaerobic conditions. No probiotic was included in this matrix plug and it was not stored at 4 °C under aerobic conditions. Scale bar = 0.5 mm.*

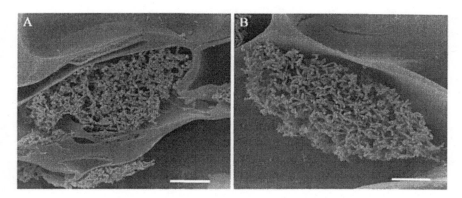

Figure 2. Growth of Lactobacillus acidophilus *(A) and* Lactobacillus reuteri *(B) in the alginic acid-calcium matrix without a probiotic. Conditions are the same as in Figure 1. Scale bars = 10 μm.*

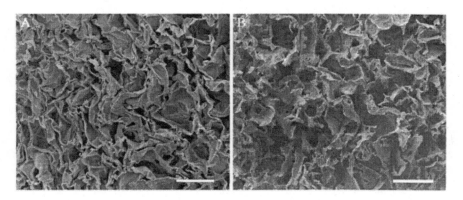

Figure 3. Structure of the alginic acid-calcium matrix when POS (A) and MCP (B) were included. Scale bars = 0.1 mm.

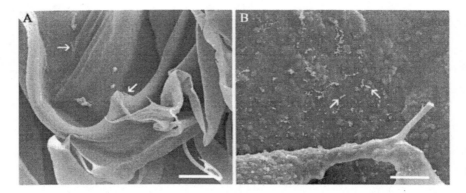

Figure 4. Growth of L. reuteri *on the POS-alginic acid-calcium matrix (A) and MCP-alginic acid-calcium matrix (B) three days after transfer of an inoculated plug to BHI broth and anaerobic conditions. These inoculated matrix plugs were stored at 4 °C under aerobic conditions for two weeks prior to transfer. Bacterial cells are labeled with arrows. Scale bars = 10 μm.*

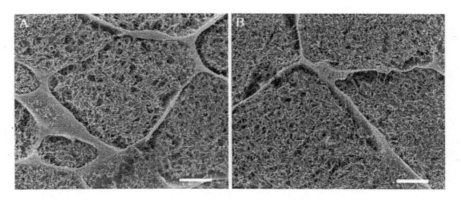

Figure 5. Growth of L. acidophilus *in the FOS-alginic acid-calcium matrix (A) and inulin-alginic acid-calcium matrix (B) one day after transfer of an inoculated plug to MRS broth and anaerobic conditions. These inoculated matrix plugs were stored at 4 °C under aerobic conditions for four weeks prior to transfer. Scale bars = 20 μm.*

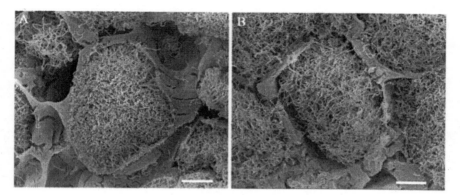

Figure 6. Growth of L. acidophilus *in the FOS-alginic acid-calcium matrix (A) and inulin-alginic acid-calcium matrix (B) two days after transfer of an inoculated plug to MRS broth and anaerobic conditions. These inoculated matrix plugs were stored at 4 °C under aerobic conditions for three weeks prior to transfer. Scale bars = 20 μm.*

Since bacterial growth was observed even when prebiotics were not included in the alginic acid-calcium matrix, and we did not quantitatively determine the amount of bacterial growth resulting from the different matrices, it is not possible to compare the prebiotic effects of the different oligo-/poly-saccharides used in these synbiotics. However, we know that POS did not have a prebiotic index equivalent to FOS until 24 hours of growth in mixed batch fecal cultures (5). Therefore, it is likely that the fructan matrices produced more rapid growth of lactobacilli compared to POS in these synbiotics. In vitro fluorescent in situ hybridization (FISH) assays using 16s rRNA probes demonstrated that POS and FOS produced significant increases in *Bifidobacteria* and *Eubacteria* while increases in lactobacilli did not reach significant levels (5). POS produced by enzymatic hydrolysis of commercial pectin supported growth of *L. acidophilus* as well as *Bifidobacteria* and prebiotic index values steadily increased during 48 hours in mixed batch fecal culture (4). Therefore, it is anticipated that a more gradual sustained prebiotic effect would be possible using POS compared to FOS in the alginic acid-calcium synbiotics. This would be an advantage for controlled release of probiotic bacteria, and the short-chain fatty acids they produce, into more distal regions of the colon. Future research will examine whether or not these hypotheses are correct.

Plant fibers have a protective effect on probiotic bacteria. Wheat dextrin was an excellent carrier for *L. rhamnosus* during freeze-drying and in chocolate-coated breakfast cereal (8), oat flour with 20% β-glucan increased the survival of this probiotic in apple juice (8) and apple slices were used on immobilize *L. casei* for lactic acid production and milk fermentation (9). We observed that the viability of lactobacilli probiotics was preserved for a month in the oligo-/poly-saccharide alginic acid-calcium matrices.

Conclusions

The alginic acid-calcium matrix provided an excellent structure for the growth of lactobacilli. This matrix is compatible with various plant oligo- and poly-saccharide prebiotics. Through careful selection of the prebiotic, probiotic bacteria can be delivered to the colon from a synbiotic matrix in a viable, actively growing state. It may be possible to direct the delivery of actively growing probiotic bacteria to different regions of the colon based on the prebiotic selected in the synbiotic matrix.

Acknowledgments

We thank Guoping Bao, Brad Shoyer and Andre White for technical assistance. The Pectasol was a generous gift from EcoNugenics. Raftilose P95 and Synergy1 were generous gifts from Orafti.

References

1. Gibson, G. R.; Roberfroid, M. B. Dietary modulation of the human colonic microbiota: introducing the concept of prebiotics. *J. Nutrit.* **1995**, *125*, 1401-1412.
2. Deardorff, J. Germs that will fight for you. Chicago Tribune http://www.venturacountystar.com/news/2007/sep/10/germs-that-will-fight-for-you/ 2007.
3. Luchansky, J. B. Use of biotherapeutics to enhance animal well being and food safety. Proceedings of the 6th International Feed Production Conference, Piacenza, Italy, November 27-28, 2000.
4. Olano-Martin, E.; Gibson, G. R.; Rastall, R. A. Comparison of the in vitro bifidogenic properties of pectins and pectic-oligosaccharides. *J. Appl. Microbiol.* **2002**, *93*, 505-511.
5. Manderson, K.; Pinart, M.; Tuohy, K. M.; Grace, W. E.; Hotchkiss, A. T.; Widmer, W.; Yadav, M. P.; Gibson, G. R.; Rastall, R. A. In vitro determination of prebiotic properties of oligosaccharides derived from an orange juice manufacturing by-product stream. *Appl. Environ. Microbiol.* **2005**, *71*, 8383-8389.
6. Eliaz, I.; Hotchkiss, A. T.; Fishman, M. L.; Rode, D. The effect of modified citrus pectin on the urinary excretion of toxic elements. *Phytother. Res.* **2006**, *20*, 859-864.

7. Liu, L.; Fishman, M. L.; Hicks, K. B. Pectin in controlled drug delivery – a review. *Cellulose* **2007**, *14*, 15-24.

8. Saarela, M.; Virkajarvi, I.; Nohynek, L.; Vaari, A.; Matto, J. Fibres as carriers for Lactobacillus rhamnosus during freeze-drying and storage in apple juice and chocolate-coated breakfast cereals. *Int. J. Food Microbiol.* **2006**, *112*, 171-178.

9. Kourkoutas, Y.; Kanellaki, M.; Koutinas, A. A. Apple pieces as immobilization support of various microorganisms. *Libens. Wisensch. Technol.* **2006**, *39*, 980-986.

Drug Delivery Systems

Chapter 6

Significant Role of Naturally Occurring Materials in Drug Delivery Technology for Tissue Regeneration Therapy

Yasuhiko Tabata

Department of Biomaterials, Field of Tissue Engineering, Institute for Frontier Medical Sciences, Kyoto University, Kyoto, Japan

As the third medical therapy following reconstructive surgery and organ transplantation, the therapy of regenerative medicine is being clinically expected. The objective of regenerative therapy is to induce the regeneration and repairing of defective and injured tissues based on the natural-healing potential of patients themselves. For successful tissue regeneration, it is important to maximize the natural cell potentials of proliferation and differentiation. For this purpose, a biomedical technology and methodology are required to create a local environment that enables cells to enhance their potentials for natural induction of tissue regeneration. This biomedical research field is called tissue engineering where the regeneration environment is built up by the naturally occurring materials of biodegradability and good bio-compatibility. When biological signaling molecules, growth factors and genes, are used, drug delivery system (DDS) is indispensable to enhance their biological activities of cell proliferation and differentiation. This paper introduces the tissue engineering application of naturally occurring materials, especially gelatin, for signaling molecules delivery to emphasize significance of DDS technology in regenerative medical therapy, briefly explaining the basic concept of tissue engineering.

Tissue Engineering Necessary for Regenerative Medical Therapy

As surgical therapies currently available, there are reconstruction surgery and organ transplantation. Although it is no doubt that these advanced therapies have saved and improved the countless lives of patients, they have clinical limitations. The reconstruction surgery almost depends on medical devices and artificial organs which cannot completely substitute the biological functions even for a single tissue or organ. In addition, the progressive deterioration of injured tissue and organ cannot be always suppressed therapeutically. One of the biggest issues for organ transplantation is the shortage of donor tissues or organs. The permanent medication of immunosuppressive agents often causes side-effects, while virus infection is not completely ruled out. In this circumstance, a new therapeutic trial, in which disease healing can be achieved based on the natural-healing potential of patients themselves, has been explored. To realize this therapy of regenerative medicine, it is necessary to provide cells a local environment suitable for their proliferation and differentiation to naturally induce cell-based tissue regeneration. It is tissue engineering that is a newly emerging biomedical engineering forms to create the regeneration environment. The objective of tissue engineering is to induce regeneration of defective or lost tissues as well as substitute the biological functions of damaged organ by maximizing the natural potentials of cells. In tissue engineering, biomaterials are used in various manners and ways to enhance the cell-mediated natural healing potential for tissue regeneration. Especially, naturally occurring materials of biodegradability and biocompatibility are preferable as biomaterials for tissue engineering applications. There are two types of tissue engineering. One is surgical tissue engineering, where biomaterials with or without cells and/or drugs combination are surgically applied to a body tissue defect to induce tissue regeneration threat for disease therapy. The other is physical tissue engineering of internal medicine. For example, drugs are physically applied to the fibrotic tissue of chronic diseases for the loosening and digestion, leading the regeneration and repairing of fibrotic disease based on the cell potential of the surrounding tissue. Both are conceptually similar from the viewpoint that disease is medically treated by making use of the natural healing potential of living body.

Fundamental Technology and Methodology of Tissue Engineering

Considering the components consisting a body tissue, there are three key factors; cells, the scaffold of cell proliferation and differentiation, and biological signaling molecules (growth factors and genes). The basic idea of regenerative medicine is that tissue regeneration is naturally induced by using the body tissue components in a single or combinational way. This process of regeneration induction is biomedically supported and promoted by the technology or methodology of tissue engineering with biomaterials. In tissue engineering, there are four fundamental technologies. Every technology greatly depends on the naturally occurring materials of in vivo degradability. The first technology is to prepare an artificial scaffold that enables cells to accelerate their proliferation and differentiation for tissue regeneration (Figure 1A). It is well recognized that the extracellular matrix (ECM) is not only a physical support of cells, but also provides a natural local environment for cell proliferation and differentiation or the cell-based morphogenesis which contributes to tissue regeneration and organogenesis (1). It is unlikely that a large-size tissue defect will be naturally regenerated and repaired only by supplying cells to the defect area. For example, one promising way is to artificially give cells a local environment suitable to induce tissue regeneration at a defect by in advance supplying the biomaterial scaffold of artificial ECM to the defect. The scaffold assists the initial attachment of cells in a homogeneous and three-dimensional manner and accelerates the subsequent proliferation and differentiation. The scaffold should be biodegradable, biocompatible to cells, and a three-dimensional structure with interconnected pores. Since the scaffold is a temporary substrate of cells to assist their proliferation and differentiation, the remaining often impairs physically the natural process of tissue regeneration. Fast degradation does not structurally support the initial step of cell-mediated tissue regeneration. The pore structure is necessary to supply oxygen and nutrients to cells present inside the scaffold and excrete the cell wastes. It is highly expected that cells residing around the scaffold of biomaterials infiltrate into the matrix and proliferate and differentiate therein if the biomaterial is biologically compatible.

When the tissue around a defect does not have any inherent potentials to regenerate, the tissue regeneration cannot be always expected only by supplying the scaffold. In such a case, the scaffold should be used combining with cells or/and biological signaling molecules (growth factors and genes) which has a natural potential to accelerate tissue regeneration. For signaling molecules, it is indispensable to contrive way how to apply the in vivo for their enhanced biological activities. It is highly expected that growth factor is required to promote tissue regeneration, if it is used efficiently. However, the direct injection of growth factor in the solution form into the site to be regenerated is generally not effective. This is because the growth factor rapidly diffused from the injected site and is enzymatically digested or deactivated. To enable the growth factor to efficiently exert its biological function, the second key technology of tissue engineering, that is drug delivery system (DDS), is required (Figure 1B). Among the DDS technologies, the controlled release of growth factor at the site of action over an extended time period is achieved by incorporating the factor into an appropriate carrier. It is also highly possible that the growth factor is protected against its proteolysis, as far as it is, at least, incorporated in the release carrier, for prolonged retention of the activity in vivo.

84

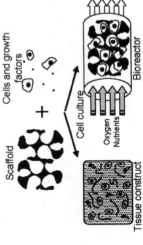

(C) Biomaterials and technology for cell culture to obtain cells clinically available.

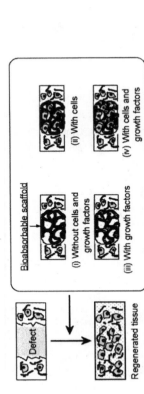

(A) Biomaterials and technology for cell scaffold to induce the in vivo regeneration of tissues and organs. The scaffold is combined to use with cells and/or growth factors depending on the site to be regenerated.

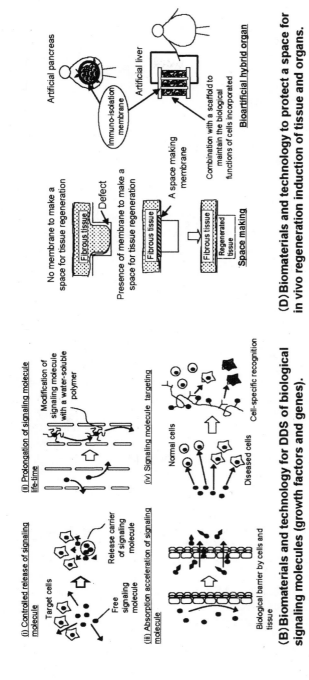

Figure 1. Role of biomaterials in tissue engineering.

The controlled release technology promotes the biological activity of growth factor for cell proliferation and consequently cell-based tissue regeneration. The release carrier should be degraded in the body since it is not needed after the growth factor release is completed. Other than the controlled release of drug, the objectives of DDS include the prolongation of drug half-life, the improvement of drug absorption, and drug targeting. For example, it is a promising approach to promote tissue regeneration by targeting a growth factor with the prolonged in vivo half-life to the tissue site to be regenerated.

It is no doubt that cells with high proliferation and differentiation potentials, so-called precursor and stem cells, are important for the therapy of regenerative medicine. However, one of the problems is the shortage of cells clinically available. Therefore, it is necessary to increase the number of stem cells with a high quality up to a level clinically applicable. To this end, the technology and methodology to isolate and culture cells are required for research and development. The scaffold described above is also important to improve the cell culture system (Figure 1C). The fourth technology is to make a space for cell-based tissue regeneration. A physical membrane of biomaterials is used to protect cells transplanted and the tissue defect to be regenerated from immunological attacks and fibroblasts or tissue infiltration, respectively (Figure1D). When a body defect is generated, the defect space is generally occupied rapidly with the fibrous tissue produced by fibroblasts which are ubiquitously present in the body and can rapidly proliferate. This is one of the typical wound healing processes to biologically maintain the living system for live save. However, once this ingrowth of fibrous tissue into the space to be regenerated takes place, the target tissue to be regenerated at the space cannot be expected any more. To prevent the tissue ingrowth, a barrier membrane to make a space necessary for tissue regeneration is highly required. One example is the immunoisolation membrane to protect cells transplanted from the biological attacks of humoral and cellular components. Thus, tissue engineering technology or methodology with biomaterials, such as cell scaffolding, space making, and DDS, is important to create a local environment of cells for their proliferation and differentiation to induce tissue regeneration.

The biomaterials used for tissue engineering are almost biodegradable because the remaining materials often become physical obstacles against tissue regeneration. Here, biodegradable materials are briefly reviewed. Table I summarizes synthetic and natural polymers of biodegradable nature. Synthetic biodegradable polymers used clinically are the homopolymer of lactic acid and its copolymers with glycolic acid and ε-caprolactone. Their biodegradable pattern can be widely and readily controlled by changing the molecular weight and copolymer compositions. One type of poly(anhydrides) and poly(cyanoacrylates) is being used as the carrier of an antitumor agent and a surgical adhesive, respectively. Other synthetic polymers have been experimentally investigated for their biomedical and pharmacological applications. Of the naturally occurring polymers, proteins (collagen, gelatin,

Table I. Biodegradable Biomaterials Applicable for Cell Scaffold

	Material	Abbreviation	Crystallinity	Typical shape
Synthetic polymer	poly(glycolide)	PGA	crystral	fabric
	poly(L-lactide-co-glycolide) (10:90)	P(L-LA/GA) (10:90)	crystral	fabric
	poly(D,L-lactide-co-glycolide) (50:50)	P(D,L-LA/GA) (50:50)	amorphous	sponge, film
	poly(D,L-lactide-co-glycolide) (85:15)	P(D,L-LA/GA) (85:15)	amorphous	fabric, sponge
	polyglycolide-co-e-caprolactone) (75:25)	P(GA/CL) (75:25)	amorphous	fabric
	poly(L-lactide)	P-L-LA	crystral	fabric
	poly(D,L-lactide)	P-D,L-LA	amorphous	sponge
	poly(L-lactide-co-e-caprolactone) (75:25)	P(L-LA/CL) (75:25)	crystral	fabric
	poly(L-lactide-co-e-caprolactone) (50:50)	P(L-LA/CL) (50:50)	amorphous	Sponge
	poly(ε-caprolactone)	PCL	crystral	Fabric
	poly(p-dioxanone)	PDS	crystral	fabric
Natural polymer	Collagen		crystral	gel, sponge
	Gelatin		amorphous	gel, sponge
	Fibrin		crystral	gel, sponge
	Polysaccharides		crystral	gel, sponge
Inorganic material	Tricalcium phosphate	TCP	crystral	porous substrate
	Calcium carbonate		crystral	porous substrate

fibrin, and albumin) and polysaccharides (chitin, hyaluronic acid, cellulose, and dextran) have been medically and pharmacologically employed. Generally, the degradation of each polymer is driven by hydrolytic and enzymatic cleavage of their main chain. Most of the synthetic polymers are fundamentally degraded by simple hydrolysis although poly(amino acids) are subjected to enzymatic degradation. On the contrary, the naturally occurring polymers are all degraded enzymatically. There is a degradation manner in which a polymer becomes water soluble as a result of the side chain elimination, disappearing from the implanted site. Metals are non-degradable materials in the body and are corroded. Ceramics are also non-biodegradable, except for tricalcium phosphate and calcium carbonate. Hydroxyapatite is not degraded in the body because of its extremely low solubility in water. Among polymers, metals, and ceramics, it is known that only the polymer shows biodegradable nature, except for a part of ceramics. Although a few synthetic polymers including poly(L-lactic acid), poly(glycolic acid), poly(ε-caprolactone), and the copolymers have been used clinically, their degradation rate are rather slow from the viewpoint of tissue engineering applications. Considering fast degradation and biocompatibility of materials, it is preferable to use naturally occurring polymers, such as protein and polysaccharide. In the following part, concrete examples of tissue regeneration with gelatin and collagen of protein are introduced to emphasize significance of naturally occurring materials in regenerative medical therapy.

Tissue engineering for clinical regenerative medicine can be classified into two categories in terms of the site where tissue regeneration or organ substitution is performed: in vitro and in vivo tissue engineering. In vitro tissue engineering involves tissue reconstruction and organ substitution which has been known as bioartificial hybrid organ. If a tissue can be reconstructed in vitro in factories or laboratories on a large scale, we can supply the tissue construct to patients when it is needed. If possible, this may be directly connected to the business of regenerative medical therapy. However, it is practically impossible to in vitro reconstruct the in vivo biological event completely only by using the basic knowledge of biology and medicine or cell culture technologies currently available. It is difficult to completely achieve in vitro tissue engineering at present, as far as it is impossible to artificially arrange a biological environment for cell-based tissue reconstruction. Another approach of in vitro tissue engineering is to substitute the biological functions of damaged organ by use of allo- or xenogeneic cells. Combination of the cells with biomaterials for immunoisolation, so-called bioartificial hybrid organ, has been investigated to substitute the function of liver and pancreas. Different from the in vitro tissue engineering, in vivo tissue engineering is advantageous from the viewpoint of the creation of local environment to induce tissue regeneration. It is likely that most biological components essential for tissue regeneration are naturally supplied by the living tissue of host. Under these circumstances, presently almost all the approaches of tissue engineering have been attempted in vivo with or without biodegradable scaffolds. This in vivo approach is more realistic to achieve the therapy of regenerative medicine and clinically acceptable if it works well. There are several examples where in vivo tissue regeneration is induced by making use of cell scaffolds or the combination with cells (2, 3).

As described above, if the tissue to be repaired has a high activity toward regeneration, the immatured cells of high proliferation and differentiation potentials infiltrate into the matrix of biodegradable scaffold implanted from the surrounding healthy tissue, resulting in formation of a new tissue. However, additional means are required if the regeneration potential of tissue is very low, because of, for instance, low concentration of cells and biological signaling molecules like growth factors responsible for new tissue generation. The simplest method is to supply a growth factor to the site of regeneration for cell proliferation and differentiation in a controllable fashion. As described above, it is undoubtedly necessary for the induction of tissue regeneration with a growth factor of in vivo instability to make use of DDS technology, for example a controlled release of the factor. Recent researches on tissue regeneration through combination of growth factors with the DDS carriers have demonstrated that a carrier is absolutely necessary to allow growth factor to exert the biological activity for in vivo tissue regeneration. However, although such significance of DDS in tissue regeneration is claimed, the controlled release of growth factor for tissue regeneration has not been always studied extensively. This is because the factor is generally costly and the DDS technology of protein is poor. It is necessary to accelerate the research and development of protein release.

In place of growth factor protein itself, recently the gene encoding growth factor has been applied to promote tissue regeneration (4, 5). For the gene-based tissue engineering, there are two future directions of basic research and clinical therapy. One is the conventional gene therapy by using plasmid DNA and viruses. When directly injected in the body, a plasmid DNA is transfected into cells around the injected site to secrete the DNA-coded protein which appears the therapeutic effect. Basically, this approach of plasmid DNA injection is one of protein therapies which can be achieved by gene-transfected cells. The angiogenesis (6) and bone tissue regeneration (7) have been attempted by use of the corresponding growth factor genes. However, there are problems to be improved for the efficiency of gene transfection and the consequent gene expression. To this end, DDS technologies for plasmid DNA are needed. The second direction is to genetically activate cells for enhanced efficacy of cell therapy. Stem cells are sometimes not powerful for cell therapy. As one trial to activate the biological activity of stem cells, it will be a promising way to genetically engineer cells for biological activation. A DDS technology or methodology assists to develop a system of non-viral gene transfection at the efficiency as high as that of viral system (8, 9). The genetically-engineered cells by adrenomedulin plasmid DNA complexed with a polysaccharide based carrier showed superior therapeutic effect to the normal cells when an animal model of cardiac infarction was treated by cell transplantation (gene-cell hybrid therapy) (10).

Tissue Regeneration by Controlled Release Technology of Biological Signaling Molecules

As described above, it is necessary for successful tissue regeneration induced by biological signaling molecules to develop the DDS technology. We have explored a biodegradable hydrogel of gelatin as the carrier matrix of signaling molecules release (*11*). One of the largest problems in protein release technology is the loss of biological activity of protein released from a protein-carrier formulation. It has been demonstrated that this activity loss mainly results from the denaturation and deactivation of protein during the formulation process. Therefore, a method to prepare the formulation of protein release with biomaterials should be exploited to minimize protein denaturation. From this viewpoint, polymer hydrogel may be a preferable candidate as the protein release carrier because of its biocompatibility and its high inertness toward protein drugs. However, it will be practically impossible to achieve the controlled release of protein over a long period of time from hydrogels which have been reported so far. This is because the protein release is generally diffusion-controlled through the water pathway present throughout the inside of hydrogels. Thus, a possible approach to tackle this problem, for example, is to immobilize a growth factor in a biodegradable hydrogel. The immobilized factor is not released by the simple diffusion, but only by the solubilization of factor in water as a result of hydrogel biodegradation. In such a carrier degradation-driven release system, the time profile of growth factor release is governed and can be changed only by that of in vivo hydrogel degradation. Chemical and physical methods are known to be available to immobilize growth factors into hydrogels. Since the former method often results in the denaturation and activity lowering of protein, it is preferable to follow the latter one in terms of activity maintenance of growth factor. Actually the physical immobilization is generally observed in the existence manner of growth factors in the natural tissue (*12*). Since some growth factors possess a positively charged site on the molecular surface, it is well recognized that they are normally stored in the body, being ionically complexed with acidic polysaccharides of ECM, such as heparan sulfate and heparin. This complexation protects the growth factor from denaturation and enzymatic degradation in vivo. The growth factor is released from the ECM as the factor complexed becomes water-soluble accompanied with polysaccharide degradation by enzymes secreted from cells according to the need. We have created the release system of growth factors which mimics that in the living body mentioned above. Figure 2 shows an idea on the controlled release of growth factor from a biodegradable hydrogel based on intermolecular interaction forces between the hydrogel polymer and growth factor. For example, a hydrogel is prepared from a biodegradable polymer with negative charges. The growth factor with a positively charged site is electrostatically interacted with the polymer chain to allow the factor to

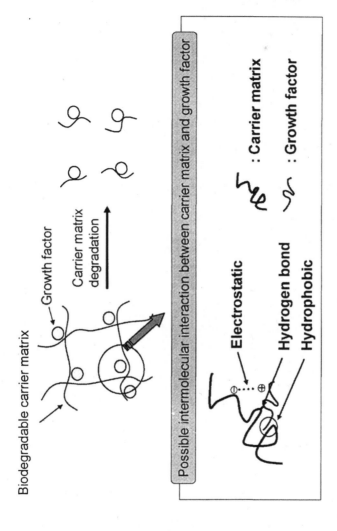

Figure 2. Schematic illustration of growth factor release followed by degradation of hydrogel carrier.

physically immobilize in the hydrogel carrier. If an environmental change, such as increased ionic strength, occurs, the immobilized growth factor will be released from the factor–carrier formulation. Even if such an environmental change does not take place, degradation of the carrier itself will also lead to growth factor release. Because the latter is more likely to happen in vivo than the former, it is preferred that the release carrier is prepared from biodegradable polymers. For growth factor release on the basis of physical interaction forces, it is necessary for the carrier material to employ a bio-safe polymer with groups able to interact with the factor. In addition, if biodegradability is required, the biomaterial to be used will be restricted to naturally occurring polymers with charged groups. Therefore, we have selected gelatin because it has physicochemical properties mentioned above and has been extensively used for industrial, pharmaceutical, and medical purposes. The biosafety of gelatin has been proved through its long clinical usage. Another unique advantage is the electrical nature of gelatin which is changed by the processing method from collagen. For example, the alkaline process of collagen results in hydrolysis of amide groups of the asparagine and glutamine residues, having a high density of carboxyl groups, which makes the gelatin negatively charged. If a growth factor to be released has the positively charged site in the molecule which interacts with acidic polysaccharides present in the ECM, the negatively charged gelatin of 'acidic' type is preferable as the carrier material. It was found that, as expected, basic fibroblast growth factor (bFGF), transforming growth factor β1 (TGFβ1) or platelet derived growth factor (PDGF) was sorbed into the acidic gelatin hydrogel mainly due to electrostatic interaction (11). Animal experiments revealed that the hydrogels prepared from the acidic gelatin were degraded in the body (13). The time profile of in vivo hydrogel degradation which can be changed by its water content was in accordance with that of growth factor retention in the hydrogel, irrespective of the hydrogel water content (Figure 3). This strongly indicates that the growth factor release is governed mainly by hydrogel degradation as described in Figure 2. It is likely that the growth factor immobilized was released from the gelatin hydrogel probably together with degraded gelatin fragments in the body as a result of hydrogel degradation. Based on this hydrogel degradation release mechanism, it is possible to achieve the controlled release of growth factor from the hydrogel of microsphere shape which has a large surface area/volume ratio (14). We have succeeded in the controlled release of a plasmid DNA by this hydrogel system (Figure 3), which enables the plasmid DNA to enhance the level of gene expression and prolong the time period of gene expression (15, 16). In addition, the hydrogel system can release not only one type of growth factor, but also two or multitypes of growth factor time or concentration-dependent fashion. For the defect of rabbit skull bone, upon applying a hydrogel incorporating either bFGF or TGFβ1 at a low dose, no bone was regenerated. However, the simultaneous release application of two factors showed synergistic effect on bone regeneration.

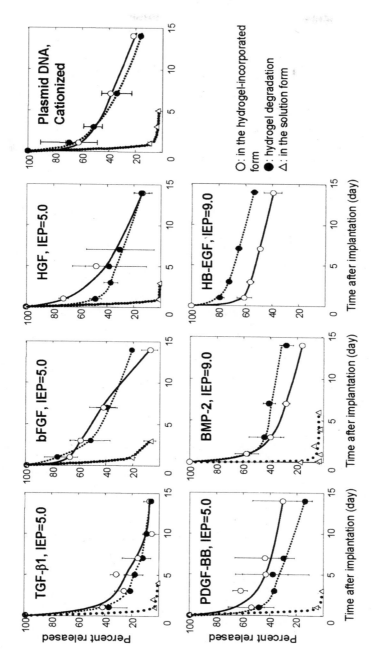

Figure 3. In vivo release profiles of biological signaling molecules from biodegradable hydrogel of various gelatins

The dual release of bFGF and hepatocyte growth factor (HGF) is superior to that of either factor for angiogenesis in terms of the dose and the biological maturation of blood vessels newly formed (2, 3, 17). The gelatin hydrogel could release the growth factor cocktail present in platelet rich plasma (PRP) to accelerate the factor –induced bone regeneration, in contrast to the PRP application without the hydrogel of release carrier (18).

bFGF was originally characterized in vitro as a growth factor for fibroblasts and capillary endothelial cells and in vivo as a potent mitogen and chemoattractant for a wide range of cells. In addition, bFGF is reported to have a variety of biological activities (19) and be effective in enhancing wound healing through induction of angiogenesis as well as regeneration of various tissues, such as bone, cartilage, and nerve. Recently, bFGF of human type in a solution form has been on the Japanese market for remedy of decubitus, a chronic ulcer of the skin caused by prolonged pressure on it, from Kaken Pharmaceutical Co. Ltd., Tokyo (product name: Fibrast® spray). When a gelatin hydrogel incorporating bFGF was subcutaneously implanted into the mouse back, significant angiogenic effect was observed around the implanted site, in marked contrast to the injection of bFGF solution at higher doses or the implantation of with bFGF-free, empty gelatin hydrogel (11). By the gelatin hydrogel system, we have succeeded in inducing the regeneration of various tissues and organs by the controlled release of various growth factors with the biological activities remaining (Table II, Figure 4).

There are two important objectives of angiogenesis in tissue engineering; the therapy of ischemic disease and in advance angiogenesis for cell transplantation. As the former example, when injected into the ischemic site of myocardial infarctio (20) or leg ischemia (21), gelatin microspheres incorporating bFGF induced angiogenesis therapeutically acceptable. This angiogenic therapy for leg ischemia has been permitted by the ethics committee of university for the clinical trial. A clinical study of bFGF-induced angiogenic therapy has been started in different hospitals to demonstrate good results.

It is no doubt that sufficient supply of nutrients and oxygen to cells transplanted into the body is indispensable for cell survival and the maintenance of biological functions. For successful cell transplantation, it is a practically promising to induce in advance angiogenesis throughout the site where cells are transplanted, by using the release system of bFGF. This technology of in advance angiogenesis efficiently improved the biological functions of pancreatic islets (22), cardiomyocytes (23), and kidney cells transplanted (24) as well as enhanced the engrafting rate of bio-artificial dermis-epidermis skin-tissue construct (25). We have succeeded in improving the cardiac functions of ischemic rat hearts by combination of cadiomyoblasts implantation with in advance angiogenesis induced by gelatin microspheres incorporating bFGF (26). These findings clearly indicate that the in advance induction of angiogenesis at the transplanted site was effective in enhancing the success rate of engrafting for cells transplanted and tissue construct prepared in vitro. The release system

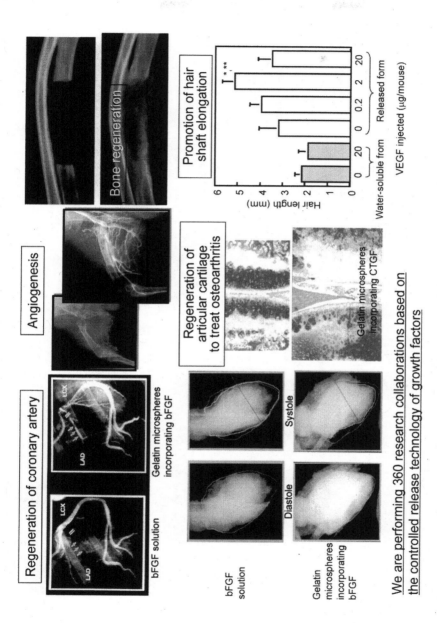

Figure 4. Controlled release technology of growth factors to realize the regeneration induction of various tissues.

Table II. Regeneration Induction of Body Tissues and Organs Based on the Controlled Release of Bioactive Growth Factors from Biodegradable Hydrogels.

Materials	Growth factor	Animal	Effect	Objective	Reference
Acidic gelatin (pI 5.0)	bFGF	Mouse, Rat, and Dog	Angiogenesis	Transplantation of Langerhans islands for diabetes therapy	(39, 40)
		Rat	"	Transplantation of hepatocytes for therapy of enzyme deficiency disease	(41)
		Rat	"	Transplantation of renal epithelial cells	(24)
		Rat and dog	"	Transplantation of cardiomyocytes	(23)
		Rat and guinea pig	"	Promoted repairing of skin dermal layer	(42)
		Rat and Pig	"	Treatment of cardiac infarction	(20, 43)
		Rabbit	"	Treatment of lower limb ischemia	(21)
		Rat, Dog, and Monkey	Osteogenesis and angiogenesis	Repairing of sternum and connective tissue	(27, 44, 45)
		Rat, Rabbit, and Monkey	Osteogenesis	Repairing of skull and long bone	(46, 47)
		Mouse	Adipogenesis	Repairing of breast and soft tissue reconstruction	(31, 48)
		Mouse	Angiogenesis and activation of hair follicle tissue	Promotion of hair growth	(49, 50)
		Dog	Periodontium repair	Repairing of periodontium	(51)
		Dog	Peripheral nerve repair	Nerve repairing	(52)
		Dog	Osteogenesis	Repairing of mandiblular bone	(53)

			Effect	Application	Ref.
	TGF-β1	Rabbit and Monkey		Repair of skull bone	(28, 54-56)
				"	
	HGF	Sheep	Chondrogenesis	Repairing of tracheal cartilages	(57)
		Mouse	Angiogenesis and activation of hair follicle tissue	Promotion of hair growth	(50)
	bFGF/TGF-β1	Rat and Pig	Angiogenesis and inhibition of apoptosis	Treatment of dilated cardiomyopathy	
		Rabbit	Osteogenesis	Repairing of skull bone	
Basic gelatin (pI 9.0)	CTGF	Rabbit	Chondrogenesis	Repairing of articular cartilage	(58)
	BMP-2	Rat, Dog, and Monkey	Osteogenesis	Repairing of skull and mandiblular bone	(59)
Collagen	TGF-β1	Dog	Chondrogenesis	Repairing of tracheal cartilages	(60)
		Rabbit	Osteogenesis	Repairing of skull bone	(61)
		Rabbit	"	Promotion of engraftment of soft tissue grafts	(62)
		Mouse	Angiogenesis and activation of hair follicle tissue	Promotion of hair growth	(50, 63)

ABBREVIATIONS: bFGF: fibroblast growth factor, TGF-β1: transforming growth factor β1, HGF: hepatocyte growth factor, CTGF: connective tissue growth factor, VEGF: vascular endothelial growth factor, BMP-2: bone morphogenetic protein 2.

enabled bFGF, TGFβ1, and BMP-2 to enhance their activity of bone renegeration (27-29) as well as bone regeneration induced by mesenchymal stem cells of bone marrow (30). For the grafting surgery of heart, the bilateral sternum artery is normally used because of the high potency. However, in spite of successful graft surgery, the sternum repairing is often delayed, much worse the infection at the resection area of sternum often takes place, while wound healing of the surrounding soft tissue is also poor due to the surgical elimination of nutrient artery. As one trial to tackle the issue, we have applied the bFGF release system to this surgical therapy because bFGF has an inherent potential to induce bone regeneration as well as angiogenesis. A hydrogel sheet incorporating bFGF was applied to the soft tissue around the sternum of diabetic rats of which sternum was cut and the bilateral arteries were ligated. As expected, bone regeneration at the cut line of sternum was achieved together with enhanced angiogenesis and the recovery of blood flow at the surrounding soft tissue (27). This bFGF-induced simultaneous regeneration of bone and the surrounding blood vessels was also observed in a clinical study. De novo adipogenesis was succeeded by the preadipocytes isolated from human fat tissues, gelatin microspheres incorporating bFGF, and a collagen sponge of cell scaffold (31). Appropriate combination of all the three materials needed to induce this adipogenesis.

This hydrogel system permits the controlled release of plasmid DNAs (Figure 3). The controlled release technology enabled a plasmid DNA and small interference RNA (siRNA) to enhance the level of gene expression and prolong the time period of gene expression (9, 16, 32, 33). We have found that the controlled release of a plasmid DNA from a biodegradable hydrogel of cationized gelatin derivative significantly enhanced the level of gene expression as well as prolong the time period expressed (16, 32). When intramuscularly injected into the ischemic leg of rats, the cationized gelatin microspheres incorporating a plasmid DNA of FGF-4 induced angiogenesis to a significantly higher extent than the plasmid DNA solution even at the dose 100 or 1000 times less than that of solution type (34). The microspheres incorporating plasmid DNA were effective in genetically activating cells and consequently enhancing the efficacy of cell therapy. Cationized microspheres incorporating the plasmid DNA of adrenomedulin were prepared to allow them to internalize into endothelial progenitor cells. Intracellular controlled release of plasmid DNA enhanced the efficiently of gene transfection to the level higher than that of adenovirus transfection. The cells genetically engineered also functioned well to achieve higher therapeutic efficacy (8). In addition to the research and development of non-viral carrier system for gene transfection, it is also necessary for enhanced gene expression of plasmid DNA to improve the culture method. For example, when cells were cultured for gene transfection on the substrate coated with plasmid DNA-cationized pullulan complex and pronectin® (reverse transfection method), the level of gene transfection and expression were significantly enhanced compared with those of the conventional transfection method where cells are incubated for gene transfection in the medium containing plasmid DNA-cationized polysaccharide complex.

Tissue Engineering of Internal Medicine Based on DDS Technology

Presently, there is no effective therapeutic strategy for chronic fibrosis diseases, such as lung fibrosis, cirrhosis, dilated cardiomyopathy, and chronic nephritis. For these diseases, the injured site of tissue and organ is normally occupied with fibrous tissue of excessive collagen fibers and fibroblasts. It is highly possible that this tissue occupation causes the physical impairment of natural healing process at the disease site. Therefore, if the fibrosis can be digested by any method to loosen or disappear, it is highly expected that the disease site is repaired based on the natural regeneration potential of the surrounding healthy tissue. It has been demonstrated that the injection of virus encoding a matrix metaloprotease (MMP) protein suppresses the tissue fibrosis to get better the disease symptoms (35). The finding strongly suggests that when collagen in the fibrous tissue is enzymatically digested, fibrosis is naturally improved or repaired due to the body potential to induce tissue regeneration which is naturally equipped in the surrounding healthy tissue. This is a new direction of tissue engineering which is defined as tissue engineering of internal medicine (Figure 5), because disease therapy induced by tissue regeneration potential is achieved by the drug treatment of internal medicine. We have demonstrated that the controlled release of a MMP-1 plasmid DNA at the medulla of chronic renal sclerosis induced the histological regeneration of kidney structure, in contrast to the plasmid DNA solution (36). When gelatin microspheres incorporating HGF were intraperitoneally injected into rats with liver cirrhosis, the liver fibrosis was histologically cured (37) (Figure 6). However, the injection of HGF solution was not effective at all and the tissue appearance was similar to that of un-treated controlled group. It is well recognized that short interference RNA (siRNA) functions to silence the activity of mRNA in the sequence-specific fashion (38). When injected into the medulla of obstruction-induced renal failure model rats, the cationized gelatin complexing the siRNA for type II receptor of TGFβ1 suppressed the renal fibrosis to a significantly great extent compared with the injection of siRNA alone (33).

Necessity of Tissue Engineering Technology in Future Regeneration Therapy

Without using precursor and stem cells with high proliferation and differentiation potentials, presently, it has been possible to induce tissue regeneration only by using the controlled release system of biological active

100

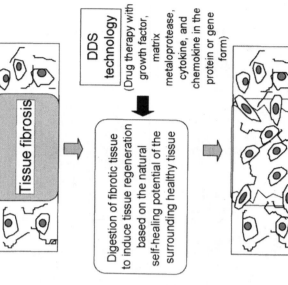

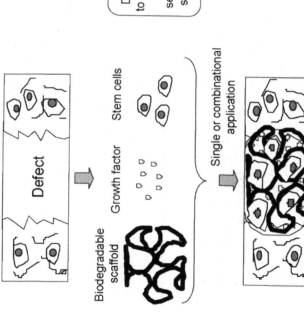

Figure 5. The concept of surgical tissue engineering and physical tissue engineering of internal medicine.

growth factors. Depending on the type of target tissue or organ and the site, it is necessary to make use of cells, their scaffold, growth factor, and the barrier membrane or their appropriate combinations. For the therapeutic approach of tissue engineering with growth factor, it is no doubt that the DDS technology or methodology is and will be indispensable in future. Although the basic idea is similar from the viewpoint of disease therapy based on the natural healing potential of patients themselves, two approaches of tissue engineering in the surgical and internal medicine manners will be extensively carried out in future. Further development of basic biology and medicine related to regenerative phenomena will accelerate the discovery of biological signaling molecules and their practical use. Under these circumstances, for both the cases, the naturally occurring polymers are important for the DDS carrier of biological signaling molecules.

If a key growth factor is supplied to the target site at the right time over the appropriate period of time and at the right concentration, we believe that the living body system will naturally direct toward the process of tissue regeneration. Once the right direction is given, it is highly possible that the intact biological system of the body starts to physiologically function, resulting in natural achievement of tissue regeneration. There is no doubt that whenever growth factors and genes are used in vivo, their combination with DDS technology is essential. However, the present technology of controlled release does not always regulate accurately the amount and time period of growth factor release. Therefore, it is practically impossible to artificially control the process of cell differentiation only by the release technology of growth factors currently available, since the differentiation process is restrictedly regulated by the complicated network of growth factor in a time, site or concentration manner.

Regenerative medical therapy, which is a new therapeutic trial based on the natural potential of living body to induce tissue regeneration with cells and tissue engineering, is the third therapy following reconstructive surgery and organ transplantation. To achieve the therapy of regenerative medicine by use of tissue engineering technology and methodology, substantial collaborative research between material, pharmaceutical, biological, and clinical scientists is needed. Even though superior stem cells with high potential of proliferation and differentiation can be obtained to use by the development of basic biology and medicine, only the transplantation of the cells will never perform their biological functions in vivo to heal the disease. In addition, it is practically impossible to directly contribute the scientific knowledge to good therapeutic results of clinical medical for patients (regenerative medical therapy) unless a local environment of cells suitable for their proliferation and differentiation is created and efficiently works properly for cells. However, one of the large problems is the absolute shortage of biomaterial researchers of tissue engineering, such as scaffolding and DDS especially release technology, aiming at tissue regeneration and the biological substitution of organ functions. Such researchers must learn medical, dental, biological, and pharmacological knowledge, in

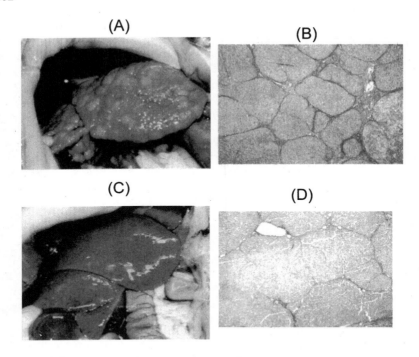

*Figure 6. Macroscopic(A,C) and Histological(B,D) appearance of rat liver
receiving 2 mg of free HGF(A,B) or gelatin microspheres incorporating
0.4 mg of HGF(C,D). (B,D; Sircoll® collagen staining).*

addition to material sciences. It is indispensable to educate the researchers of interdisciplinary field who not only have an engineering background, but also can understand basic biology and medicine or clinical medicine necessary for research and development of tissue engineering. One of the representative interdisciplinary research fields is DDS. The DDS technology is also applicable to create the non-viral vectors to prepare genetically-engineered cells for cell transplantation-based regenerative medicine. Research and development of non-viral vectors with a high efficiency of gene transfection for stem cells are highly required. Tissue engineering technology can be used surgically to the tissue defect for regeneration induction thereat, and applied to newly develop a therapeutic method for chronic fibrosis diseases by making use of methodology of internal medicine.

As tissue engineering is still in its infancy, it will take a long time to become well established although a part of the research projects has already come close to the stage of clinical applications. Increasing significance of drug delivery in future will further help progress of tissue engineering. We will be happy if this short review stimulates readers' interest in the idea and research field of tissue engineering to assist understanding of release technology importance in tissue engineering.

References

1. Ohlstein, B.; Kai, T.; Decotto, E.; Spradling, A. *Curr Opin Cell Biol* **2004**, *16*(6), 693-699.
2. Tabata, Y. *Drug Discov Today* **2005**, *10*(23-24), 1639-4166.
3. Tabata, Y., Tissue regeneration based on growth factor release. Tissue Eng **2003**, 9 Suppl 1, S5-15.
4. Polak, J.; Hench, L. *Gene Ther* **2005**, *12*(24), 1725-1733.
5. Alessandri, G.; Emanueli, C.; Madeddu, P., *Ann N Y Acad Sci* **2004**, *1015*, 271-284.
6. Lee, J. S.; Feldman, A. M. *Nat Med* **1998**, *4*(6), 739-742.
7. Bonadio, J. *Ann N Y Acad Sci* **2002**, *961*, 58-60.
8. Nagaya, N.; Kangawa, K.; Kanda, M.; Uematsu, M.; Horio, T.; Fukuyama, N.; Hino, J.; Harada-Shiba, M.; Okumura, H.; Tabata, Y.; Mochizuki, N.; Chiba, Y.; Nishioka, K.; Miyatake, K.; Asahara, T.; Hara, H.; Mori, H. *Circulation* **2003**, *108*(7), 889-895.
9. Yamamoto, M.; Tabata, Y. *Adv Drug Deliv Rev* **2006**, *58*(4), 535-554.
10. Jo, J.; Nagaya, N.; Miyahara, Y.; Kataoka, M.; Shiba-Harada, M.; Kangawa, K.; Tabata, Y. *Tissue Engineering* **2006**, in press.
11. Ikada, Y.; Tabata, Y. Adv Drug Deliv Rev **1998**, *31*(3), 287-301.
12. Taipale, J.; Keski-Oja, J. *Faseb J.* **1997**, *11*(1), 51-59.
13. Tabata, Y.; Nagano, A.; Ikada, Y. *Tissue Eng* **1999**, *5*(2), 127-138.
14. Tabata, Y.; Hijikata, S.; Muniruzzaman, M.; Ikada, Y. *J Biomater Sci Polym Ed* **1999**, *10*(1), 79-94.
15. Fukunaka, Y.; Iwanaga, K.; Morimoto, K.; Kakemi, M.; Tabata, Y. *J Control Release* **2002**, *80*(1-3), 333-343.
16. Kushibiki, T.; Tabata, Y. Curr Drug Deliv. **2004**, *1*(2), 153-163.
17. Marui, A.; Kanematsu, A.; Yamahara, K.; Doi, K.; Kushibiki, T.; Yamamoto, M.; Itoh, H.; Ikeda, T.; Tabata, Y.; Komeda, M. *J Vasc Surg* **2005**, *41*(1), 82-90.
18. Hokugo, A.; Ozeki, M.; Kawakami, O.; Sugimoto, K.; Mushimoto, K.; Morita, S.; Tabata, Y. *Tissue Eng* **2005**, *11*(7-8), 1224-1233.
19. Rifkin, D. B.; Moscatelli, D. *J Cell Biol* **1989**, *109*(1), 1-6.
20. Iwakura, A.; Fujita, M.; Kataoka, K.; Tambara, K.; Sakakibara, Y.; Komeda, M.; Tabata, Y. *Heart Vessels* **2003**, *18*(2), 93-99.
21. Nakajima, H.; Sakakibara, Y.; Tambara, K.; Iwakura, A.; Doi, K.; Marui, A.; Ueyama, K.; Ikeda, T.; Tabata, Y.; Komeda, M. *J Artif Organs* **2004**, *7*(2), 58-61.
22. Balamurugan, A. N.; Gu, Y.; Tabata, Y.; Miyamoto, M.; Cui, W.; Hori, H.; Satake, A.; Nagata, N.; Wang, W.; Inoue, K. *Pancreas* **2003**, *26*(3), 279-285.

104

23. Sakakibara, Y.; Nishimura, K.; Tambara, K.; Yamamoto, M.; Lu, F.; Tabata, Y.; Komeda, M. *J Thorac Cardiovasc Surg* **2002**, *124*(1), 50-56.

24. Saito, A.; Kazama, J. J.; Iino, N.; Cho, K.; Sato, N.; Yamazaki, H.; Oyama, Y.; Takeda, T.; Orlando, R. A.; Shimizu, F.; Tabata, Y.; Gejyo, F. *J Am Soc Nephrol* **2003**, *14*(8), 2025-2032.

25. Tsuji-Saso, Y.; Kawazoe, T.; Morimoto, N.; Tabata, Y.; Taira, T.; Tomihata, K.; Utani, A.; Suzuki, S. *Scand J Plast Reconstr Surg Hand Surg* **2007**, in press.

26. Tambara, K.; Premaratne, G. U.; Sakaguchi, G.; Kanemitsu, N.; Lin, X.; Nakajima, H.; Sakakibara, Y.; Kimura, Y.; Yamamoto, M.; Tabata, Y.; Ikeda, T.; Komeda, M. Circulation **2005**, *112*, (9), I129-I134.

27. Iwakura, A.; Tabata, Y.; Tamura, N.; Doi, K.; Nishimura, K.; Nakamura, T.; Shimizu, Y.; Fujita, M.; Komeda, M. *Circulation* **2001**, *104*(12 Suppl 1), I325-I329.

28. Yamamoto, M.; Tabata, Y.; Hong, L.; Miyamoto, S.; Hashimoto, N.; Ikada, Y. *J Control Release* **2000**, *64*(1-3), 133-142.

29. Yamamoto, M.; Takahashi, Y.; Tabata, Y. *Biomaterials* **2003**, *24*(24), 4375-4383.

30. Tabata, Y.; Hong, L.; Miyamoto, S.; Miyao, M.; Hashimoto, N.; Ikada, Y. *J Biomater Sci Polym Ed* **2000**, *11*(8), 891-901.

31. Kimura, Y.; Ozeki, M.; Inamoto, T.; Tabata, Y. *Biomaterials* **2003**, *24*(14), 2513-2521.

32. Kushibiki, T.; Tomoshige, R.; Fukunaka, Y.; Kakemi, M.; Tabata, Y. *J Control Release* **2003**, *90*(2), 207-216.

33. Kushibiki, T.; Nagata-Nakajima, N.; Sugai, M.; Shimizu, A.; Tabata, Y. *J Control Release* **2006**, *110*(3), 610-617.

34. Kasahara, H.; Tanaka, E.; Fukuyama, N.; Sato, E.; Sakamoto, H.; Tabata, Y.; Ando, K.; Iseki, H.; Shinozaki, Y.; Kimura, K.; Kuwabara, E.; Koide, S.; Nakazawa, H.; Mori, H. J Am Coll Cardiol **2003**, *41*(6), 1056-1062.

35. Iimuro, Y.; Nishio, T.; Morimoto, T.; Nitta, T.; Stefanovic, B.; Choi, S. K.; Brenner, D. A.; Yamaoka, Y. *Gastroenterology* **2003**, *124*(2), 445-458.

36. Aoyama, T.; Yamamoto, S.; Kanematsu, A.; Ogawa, O.; Tabata, Y. *Tissue Eng* **2003**, *9*(6), 1289-1299.

37. Oe, S.; Fukunaka, Y.; Hirose, T.; Yamaoka, Y.; Tabata, Y. *J Control Release* **2003**, *88*(2), 193-200.

38. Brummelkamp, T. R.; Bernards, R.; Agami, R. *Science* **2002**, *296*(5567), 550-553.

39. Wang, W.; Gu, Y.; Tabata, Y.; Miyamoto, M.; Hori, H.; Nagata, N.; Touma, M.; Balamurugan, A. N.; Kawakami, Y.; Nozawa, M.; Inoue, K. *Transplantation* **2002**, *73*(1), 122-129.

40. Sakurai, T.; Satake, A.; Sumi, S.; Inoue, K.; Nagata, N.; Tabata, Y.; Miyakoshi, J. *Pancreas* **2004**, *28*(3), e70-79.

41. Ogawa, K.; Asonuma, K.; Inomata, Y.; Kim, I.; Ikada, Y.; Tabata, Y.; Tanaka, K. *Cell Transplant* **2001**, *10*(8), 723-729.

42. Kawai, K.; Suzuki, S.; Tabata, Y.; Ikada, Y.; Nishimura, Y. *Biomaterials* **2000**, *21*(5), 489-499.
43. Sakakibara, Y.; Tambara, K.; Sakaguchi, G.; Lu, F.; Yamamoto, M.; Nishimura, K.; Tabata, Y.; Komeda, M. *Eur J Cardiothorac Surg* **2003**, *24*(1), 105-111; discussion 112.
44. Iwakura, A.; Tabata, Y.; Miyao, M.; Ozeki, M.; Tamura, N.; Ikai, A.; Nishimura, K.; Nakamura, T.; Shimizu, Y.; Fujita, M.; Komeda, M., *Circulation* **2000**, *102*(19 Suppl 3), III307-311.
45. Iwakura, A.; Tabata, Y.; Koyama, T.; Doi, K.; Nishimura, K.; Kataoka, K.; Fujita, M.; Komeda, M. *J Thorac Cardiovasc Surg* **2003**, *126*(4), 1113-1120.
46. Yamada, K.; Tabata, Y.; Yamamoto, K.; Miyamoto, S.; Nagata, I.; Kikuchi, H.; Ikada, Y. *J Neurosurg* **1997**, *86*(5), 871-875.
47. Tabata, Y.; Yamada, K.; Hong, L.; Miyamoto, S.; Hashimoto, N.; Ikada, Y. J Neurosurg **1999**, *91*, (5), 851-856.
48. Hiraoka, Y.; Yamashiro, H.; Yasuda, K.; Kimura, Y.; Inamoto, T.; Tabata, Y. *Tissue Eng* **2006**, *12*(6), 1475-1487.
49. Ozeki, M.; Tabata, Y. *Tissue Eng* **2002**, *8*(3), 359-366.
50. Ozeki, M.; Tabata, Y. *Biomaterials* **2003**, *24*(13), 2387-2394.
51. Nakahara, T.; Nakamura, T.; Kobayashi, E.; Inoue, M.; Shigeno, K.; Tabata, Y.; Eto, K.; Shimizu, Y. *Tissue Eng* **2003**, *9*(1), 153-162.
52. Mligiliche, N. L.; Tabata, Y.; Ide, C. *East Afr Med J* **1999**, *76*(7), 400-406.
53. Yokota, S.; Kinoshita, Y.; Fukuoka, S.; Mizunuma, H.; Ozono, S.; Tabata, Y.; Ikada, Y. *J Jpn Stomatol Soc* **2002**, *51*, 324-334.
54. Hong, L.; Tabata, Y.; Miyamoto, S.; Yamada, K.; Aoyama, I.; Tamura, M.; Hashimoto, N.; Ikada, Y. *Tissue Eng* **2000**, *6*(4), 331-340.
55. Hong, L.; Miyamoto, S.; Hashimoto, N.; Tabata, Y. *J Biomater Sci Polym Ed* **2000**, *11*(12), 1357-1369.
56. Hong, L.; Tabata, Y.; Miyamoto, S.; Yamamoto, M.; Yamada, K.; Hashimoto, N.; Ikada, Y. *J Neurosurg* **2000**, *92*(2), 315-325.
57. Kojima, K.; Ignotz, R. A.; Kushibiki, T.; Tinsley, K. W.; Tabata, Y.; Vacanti, C. A. *J Thorac Cardiovasc Surg* **2004**, *128*(1), 147-153.
58. Nishida, T.; Kubota, S.; Kojima, S.; Kuboki, T.; Nakao, K.; Kushibiki, T.; Tabata, Y.; Takigawa, M. *J Bone Miner Res* **2004**, *19*(8), 1308-1319.
59. Hong, L.; Tabata, Y.; Yamamoto, M.; Miyamoto, S.; Yamada, K.; Hashimoto, N.; Ikada, Y. *J Biomater Sci Polym Ed* **1998**, *9*(9), 1001-1014.
60. Okamoto, T.; Yamamoto, Y.; Gotoh, M.; Huang, C. L.; Nakamura, T.; Shimizu, Y.; Tabata, Y.; Yokomise, H. *J Thorac Cardiovasc Surg* **2004**, *127*(2), 329-334.
61. Ueda, H.; Hong, L.; Yamamoto, M.; Shigeno, K.; Inoue, M.; Toba, T.; Yoshitani, M.; Nakamura, T.; Tabata, Y.; Shimizu, Y. *Biomaterials* **2002**, *23*(4), 1003-1010.
62. Tabata, Y.; Miyao, M.; Ozeki, M.; Ikada, Y. *J Biomater Sci Polym Ed* **2000**, *11*(9), 915-930.
63. Ozeki, M.; Tabata, Y. *Biomaterials* **2002**, *23*(11), 2367-2373.

Chapter 7

pH and Temperature Sensitive Block Copolymer Hydrogels Based on Poly(ethylene glycol) and Sulfamethazine

Doo Sung Lee, Woo Sun Shim, and Dai Phu Huynh

Department of Polymer Science and Engineering, SungKyunKwan University, Suwon, Gyeonggi 440–746, Korea

A novel pH and temperature sensitive block copolymer was prepared by adding a pH sensitive moiety to a temperature sensitive block copolymer. An oligomer incorporating pH-sensitive moieties and containing an s-triazine ring was synthesized from sulfamethazine (SM) by radical polymerization. Novel pH- and temperature-sensitive bio-degradable penta block copolymers were synthesized from the sulfamethazine oligomer (OSM) incorporating pH-sensitive moieties and a temperature sensitive degradable triblock copolymer hydrogel style ABA based on PEG. The sol-gel phase transition behavior of these block copolymers was investigated both in solution and injection into PBS buffer at pH 7.4 and 37°C. Aqueous solutions of these block copolymers showed sol-gel transition behavior upon both temperature and pH changes under physiological conditions (37°C, pH 7.4). When the block copolymer solution is in the sol state at 10°C and pH 8.0, both temperature and pH changes are needed for gelation to occur. The sol-gel transition properties of these block copolymers are influenced by the hydrophobic/hydrophilic balance of the copolymers, block length, hydrophobicity, stereoregularity of the hydrophobic components within the block copolymer, and the ionization of the pH functional groups in the copolymer, which depends on the environmental pH. The degradation time of these polymers could be controlled by changing the hydrophobic block of the temperature sensitive degradable triblock copolymer hydrogel. These materials could be employed as injectable carriers for hydrophobic drugs and proteins, etc. Gelation inside the needle can be prevented by increasing the temperature during the injection of the hydrogel, because it does not change into the gel form solely upon increasing the temperature. This material could even be used for a long guide catheter into the body.

Hydrogels that exhibit both liquid-like and solid-like behavior exhibit a wide variety of functional properties (i.e. swelling, mechanical, permeation, surface and optical), and these have provided many potential applications for hydrogels in fields such as medicine, agriculture, and biotechnology *(1-4)*. Stimuli-sensitive hydrogels have attracted considerable attention as intelligent materials in the fields of biochemistry and biomedicine, due to their ability to detect environmental changes and undergo structural changes by themselves, such as changes in their solubility and swelling ratio *(5-8)*. Various stimuli-sensitive hydrogels that respond to pH *(9)*, temperature *(10, 11)*, electric fields *(12, 13)*, and other stimuli have been studied both experimentally and theoretically *(14, 15)*. Among the stimuli-sensitive materials that have been developed so far, polymers showing a sol-to-gel transition with changing temperature have been proposed for use as injectable drug delivery systems *(16-18)*. For example, temperature responsive hydrogels (e.g., Pluronics (BASF), Poloxamers (ICI) *(19)* and block copolymers composed of poly(ethylene oxide) (PEO) and polypropylene oxide (PPO) *(20)* have been studied by many researchers. In aqueous solutions, these polymers undergo a temperature induced reversible sol-gel transition upon heating and cooling. Neither Pluronics nor Poloxamers are considered an optimal system for the delivery of drugs, because they are nonbiodegradable. Therefore, biodegradable thermo-reversible hydrogels have been studied as controlled release drug carriers, because of their nontoxicity and biocompatibility. Block copolymers composed of PEG (poly(ethylene glycol)) and PLA (poly(lactic acid)) [or PLGA (poly(lactic acid-*co*-glycolic acid))] *(21)* chitosan derivatives *(22)*, polyphosphazene *(23)* methylcellulose *(24)*, etc., have been proposed as biodegradable thermo-sensitive hydrogels. Among these biodegradable thermo-sensitive hydrogels, PLGA-PEGPLGA hydrogel has been used as an injectable drug delivery system, owing to its long-term persistence in the gel form *(25)*. However, these hydrogels have several unresolved drawbacks, which limit their application in injectable drug delivery systems. When temperature-sensitive hydrogels are injected into the body with a syringe, they tend to change into a gel as the needle becomes warmed by the body temperature. This change makes it difficult to inject them into the body. Also, after being injected, these hydrogels tend to undergo rapid degradation of the block copolymer, which consequently produces the acidic monomers such as lactic acid or glycolic acid. Since the low-pH environment of the hydrogel caused by the acidic monomer is known to be deleterious to some proteins and nucleic acids, the pH change which occurs within these biodegradable hydrogels is an important consideration *(26)*.

Kim et al. used two crosslinked copolymers, poly(methacrylic acid-co-methacryloxyethyl glucoside) [P(MMA-co-MEG)] and poly(methacrylic acid-g-ethylene glycol) [P(MMA-g-EG)], to determine the mechanism of penetrant transport through anionic pH sensitive hydrogels *(27)*. They observed that the water transport mechanism was significantly dependent on the pH of the swelling medium. At high pH (higher than the pKa of the gel), the water transport was controlled more by polymer relaxation than by penetrant diffusion. For both P(MMA-co-MEG) and P(MMA-g-EG) hydrogels, the swelling mechanism exhibited little dependence on the copolymer compositions of the hydrogels at the same pH. As such, the characteristics of these systems for drug delivery applications were investigated *(28)*, and it was found that the mesh size of these hydrogels changed from small (18-35 Å) in the collapsed state at pH 2.2 to very large (70-111 Å) at pH 7.0, and increased between two and six times during the swelling process, demonstrating some potential disadvantages for use as drug delivery systems by the subcutaneous injection method.

Herein, we report investigations of novel pH and temperature sensitive block copolymer hydrogels based on polyethylene glycol (PEG) and sulfamethazine *(29, 30)*. Sulfamethazine oligomer (OSM) was used as a pH-sensitive moiety. The two temperature sensitive block copolymers used were an ABA type block copolymer composed of poly(ε-caprolactone-*co*-lactide) (PCLA) and PEG (PCLA-PEG-PCLA) and block copolymer polyethylene glycol–poly (glycolide-co-ε-caprolactone) (PCGA-PEG-PCGA), which have different degradation times depending on PCLA and PCGA, respectively. The objectives of this research were to study the effects of OSM, PCLA, PCGA, and PEG on the sol-gel phase transition of these pH/temperature sensitive penta-block hydrogels, to study the degradation of these block copolymers, and to study drug loading into polymer solutions and release out of the hydrogels of these copolymers.

Experimental Details

Materials and Methods

Poly(ethylene glycol) (PEG) was purchased from Sigma-Aldrich (St. Louis, MO) (Mn=1000, 1500 and 2000) and ID Biochem, Inc. (Seoul, Korea)

(Mn=1750), and recrystallized in n-hexane and dried in vacuum for 3 days prior to use. D,L-lactide (LA), ε-caprolactone (CL), stannous octoate [Sn(Oct)$_2$] were obtained from Sigma and dried for 24 hours under vacuum at ambient temperature prior to use. N,N –dimethyl formamide (DMF) anhydrous and glycolide (GA) were obtained from Polyscience Boehringer Ingelheimm. Methylene chloride (anhydrous), methacryloyl chloride, 3-mercaptopropionic acid (MPA), dicyclohexyl carboimide (DCC), and 4-dimethyl amino pyridine (DMAP) were used as received from Aldrich, whereas, sulfamethazine and 2, 2'-azobisisobutyronitrile (AIBN) were supplied by Sigma and Junsei Co., respectively. AIBN was recrystallized from methanol twice prior to use. Sulfamethazine was obtained from Sigma and used as received. Chloroform (CDCl$_3$) and diethyl ether were both obtained from Samchun, while paclitaxel (PTX) was purchased from Samyang Genex Corporation. All other reagents were of analytical grade and used without further purification.

Temperature-Sensitive Block Copolymer (PCLA-PEG-PCLA) and (PCGA-PEG-PCGA)

The synthesis of the PCLA-PEG-PCLA and PCGA-PGE-PCGA block copolymers was performed through a ring-opening copolymerization reaction using PEG as an initiator and Sn(Oct)$_2$ as a catalyst. The ratios of PEG/PCLA and CL/LA, PEG/PCGA and CL/GA were adjusted by altering the feed ratios of PEG, CL, LA and GA. The detailed synthesis was as follows: PEG and Sn(Oct)$_2$ were added to a two-neck round-bottom flask and were dried for 4 h under vacuum at 110°C. After cooling the flask to room temperature, LA and CL or GA and CL were added under dry nitrogen, and the reactant mixture was dried for 1 h under vacuum at 60°C. Then, the temperature was raised slowly to 130°C, and the reaction was performed over a period of 24 h under dry nitrogen. The reactants were then cooled to room temperature, dissolved in methylene chloride (MC), and added to excess diethyl ether, causing the products to precipitate. The precipitated block copolymer was then dried under vacuum at 40°C over 48 h, affording a yield of over 75%.

Sulfamethazine Oligomer

The sulfamethazine monomer (SMM) was synthesized from sulfamethazine (SM) and methacryloyl chloride. First, SM (0.1 mol) and sodium hydroxide (0.1 mol) were dissolved in aqueous acetone (100 mL, 1:1 v/v), and methacryloyl chloride (0.12 mol) was then added dropwise to the solution with stirring (0°C). The resulting mixture was then stirred for a further 3 h at 0°C. The precipitated

SMM was filtered from the solution, washed with distilled water, and then dried under vacuum at room temperature for 48 h. The SMM yield was approximately 85% after drying.

The sulfamethazine oligomer (OSM) containing a carboxyl acid end group was synthesized by conventional radical polymerization with SMM, AIBN, and 3-mercaptopropionic acid (MPA). The molecular weight of the OSM was controlled by altering the feed ratios of AIBN and MPA. The synthetic procedure was as follows: SMM (90 mmol) was dissolved in anhydrous N,N-dimethylformamide (DMF) (150 mL), after which AIBN (9 mmol) and MPA (9 mmol) were added under dry nitrogen to afford an SMM/AIBN/MPA mole ratio of 100/10/10. The temperature of the reactants was slowly increased to 85°C, and the reaction was carried out for 48 h. Subsequently, after evaporating the solvent (DMF), the resultant was redissolved in tetrahydrofuran (THF). The slow addition of the THF solution to excess diethyl ether resulted in the precipitation of OSM, which was filtered and dried slowly at 40°C for 48 h. The yield (ratio of the weight of the final product (OSM) to the total feed amount of SMM and MPA) was 90%.

Coupling of Sulfamethazine Oligomer with Temperature-Sensitive Block Copolymers

The temperature-sensitive triblock copolymer and OSM were coupled together using 1, 3-dicyclohexylcarbodiimide (DCC) and 4-(dimethylamino) pyridine (DMAP) as a catalyst. The coupling reaction process was as follows: The tri block copolymer (0.1mol) was weighed into a two-neck flask and dried under vacuum at 85°C for 2 h. OSM (0.24 mol) was then added to the flask under dry nitrogen, and the reactant mixture was dried under vacuum at 85°C for 1 h in order to completely remove any moisture. The reactant mixture was cooled to room temperature under dry nitrogen and, then, an anhydrous MC solution (60 mL) containing DCC and DMAP was added to the flask using a glass syringe to afford a triblock copolymer/OSM/DCC/DMAP feed ratio of 1/2.4/2.8/0.28 mol. The reaction was carried out at room temperature for 48 h. Although OSM is insoluble in MC, it reacts with the triblock copolymer due to the high solubility of the triblock copolymer in MC. Over the course of the coupling reaction, DCC was slowly converted into dicyclohexylurea (DCU). The residual DCC was also converted into DCU by the addition of two or three drops of water, and the combined DCU byproducts were precipitated and removed (0.4 μm filter paper) along with the residual OSM. The final product was obtained by pouring the filtered reactant mixture into excess diethyl ether, and the resulting precipitate was dried under vacuum at 40°C for more than 48 h to give a final yield of over 60%.

Characterization

The number-average molecular weight (M_n) and molecular distribution (MWD) of the as-synthesized block copolymers were determined by gel permeation chromatography (GPC) measurement on a Waters Model 410, equipped with 4 μm-styragel columns from 500 to 10 Å in series, at a flow rate of 1.0 ml/min (eluent: THF, 36°C, PEG as standard). ^{1}H-NMR measurements were performed on a Varian Unity Inova 500 instrument (500 MHz) to determine the molecular structures and compositions of PEG, CL, and GA (*30, 31*).

Phase Diagram Measurements

The block copolymers were dissolved at a given concentration in a buffer solution (in a 4 mL vial) for 1 day at 0°C. The buffer solution was prepared using PBS tablets and NaOH (0.9 wt %). The pH of the block copolymer solution was adjusted to a specific pH by adding small amounts of 5 M HCl solution at 2°C. Each solution was kept at 4°C for 30 min in a water bath. The vial was then slowly heated in a water bath in intervals of 2°C. The vial was held at each temperature for 10 min to allow it to equilibrate and then laid down horizontally for a further 1 min. The sol (flow)-gel (no flow) phase-transition temperature of the block copolymer solutions was determined using this method. The measurements were repeated three times, and each point represented an average with an accuracy of 2°C.

Sol-gel Transition by In Vitro Test

The penta-block copolymers were dissolved to obtain a solution at a concentration of 20 wt%, at pH 8.0 and 10°C. Then these solutions were injected into a 30 mL vial containing 20 ml of PBS buffer (pH 7.4 and 8.0) at 37°C through a spiral glass. Following this, the vials were shaken to check the ol-gel state of the sample.

Degradation of Block Copolymer

The degradability of the block copolymers was determined by following the changes in the molecular weight over time. The block copolymer solution was prepared at 0°C using a method similar to that used for the phase diagram experiment and was adjusted to pH 7.4. 0.5 g of the triblock and penta-block solutions at 20 wt % in water (pH 7.4) were placed in a 4 ml vial and incubated at 37°C for 30 minutes. PBS buffer (3 ml) at pH 7.4 and 37°C was added to the solution. Samples were then taken at designated time intervals and freeze-dried. The change in the molecular weight was determined by GPC.

Cytotoxicity Determination

The cytotoxicity was characterized as a decrease in the metabolic rate measured using the XTT (2,3-bis(2-methoxy-4-nitro-5-susfophenyl)-2H-tetrazolium-5-carboxanilide) assay *(31, 32)*. Cells were plated in 96-well plates at an initial density of 10,000 cells/well in 100 μL of growth medium (90% Dulbecco's modified Eagle's medium, 10% fetal bovine serum, 4500 mg/L glucose L-glutamine, 110 mg/L sodium pyruvate and sodium bicarbonate). The cells were grown for 24h, after which the growth medium was removed and replaced with fresh, serum-free medium containing the polymer. The cells was incubated with the polymer for 4 h at 37°C, and the medium was replaced with complete growth medium for 24 h. XTT labeling mixture was prepared by mixing XTT labeling agent (50 μL) and an electron coupling agent (1 μL) and 50 μL of the XTT labeling mixture was added to each well. The samples were incubated for 4 h at 37°C under 5 % CO_2 and the absorbance was read between 492 nm and 690 nm.

Drug Loading and Releasing Experiment

The drug paclitaxel (PTX) was loaded into the penta-block copolymer solutions (20 wt % in water at pH 8.0) at 0°C over a period of 1 day. The sample pH was adjusted to pH 7.4 with sodium hydroxide (5 M) and HCl (5 M) and the solution maintained at 0°C for 12 hours. Subsequently, 0.5 g of the mixture was placed in a 4 ml vial and incubated at 37°C for 30 minutes. Fresh serum (3 ml, 2.4 wt % Tween 80, 4 wt % Cremophor EL in PBS buffer at pH 7.4) at 37°C was added to the vial samples. At a given time, 1.5 ml of the serum was extracted from the vial sample and freeze-dried. The amount of paclitaxel in the samples was determined by HPLC (Column: C18, 250 × 4.0 mm, 5.0 μm; mobile phase: ACN/H_2O = 2/8; flowrate: 0.5 mL/min; detector: UV at 254 nm).

Results and Discussion

Synthesis and Characterization

The molecular structures of the synthesized SMM and OSM were confirmed by ^{1}H NMR, as shown by the corresponding spectra and peak assignments in

Figure 1. The aromatic (c, d) and amine (e) protons shown at 7.65 ppm (c), 6.55 ppm (d), and 5.98 ppm (e) in the sulfamethazine (SM) spectrum (A) were observed to shift to 7.95 ppm (c), 7.85 ppm (d), and 10.08 ppm (e) in the corresponding SMM spectrum (B), respectively. In confirming the formation of OSM, the methyl signal (1.98 ppm, f) in the SMM spectrum is shifted upfield to 1.10 ppm (f') in spectrum (C), while the corresponding ethylene signals (=CH2, g, h) show a significant decrease in intensity. The peak of the hydrogen proton in the sulfonamide group (SO$_2$NH) is shown at 11.6 ppm in the ^{1}H NMR spectra for all three samples. These ^{1}H NMR results show conclusively that SMM was successfully synthesized. On the other hand, further evidence is required to confirm the molecular structure of OSM. The molecular weight of OSM was controlled by adjusting the feed ratio of the monomer, the initiator, the transfer agent, and the reaction time. Table I shows the molecular weight of OSM, as determined by GPC (relative to PEG standards).

Table I. Sulfamethazine Oligomers

Feed Ratio (mol ratio)[a]	Reaction Time (h)	M_n[b]	M_w/M_n[b]
1/0.1/0.1 (OSM$_1$)	48	1144	1.35
1/0.1/0.2 (OSM$_2$)	48	937	1.24
1/0.1/0.1 (OSM$_3$)	40	904	1.32
1/0.1/0.2 (OSM$_4$)	40	806	1.26

[a][monomer]/[initiator]/[transfer agent]. [b] Measured by GPC relative to PEG standards.

Various PCGA-PEG-PCGA block copolymers were obtained by ring opening polymerization. The number average molecular weight (Mn) of the block copolymers can be calculated by comparison of the peak ratios of CL and GA with those of PEG (of known molecular weight) in the ^{1}H-NMR spectrum. Figure 2 shows a representative ^{1}H-NMR spectrum of the PCGA-PEG-PCGA block copolymer and its chemical structure. The characteristic signal appearing at 3.6 ppm was assigned to the methylene protons of the EO units and those at the ends of the CL units, while the signals at 4.68 and 2.35 ppm correspond specifically to the methylene protons of the GA unit and those at the beginning of the CL unit, respectively. The molecular compositions of the synthesized block copolymers were obtained by calculating the corresponding peak areas (33).

Various PCLA-PEG-PCLA block copolymers were obtained from the ring-opening polymerization reaction. The number average molecular weight (*M*n) of the block copolymers was calculated using the ^{1}H NMR spectrum of a PEG standard of known molecular weight. Figure 3 shows the representative 1H NMR spectrum of the PCLA-PEG-PCLA block copolymer and its chemical structure. All of the proton signals of the block copolymer were assigned as

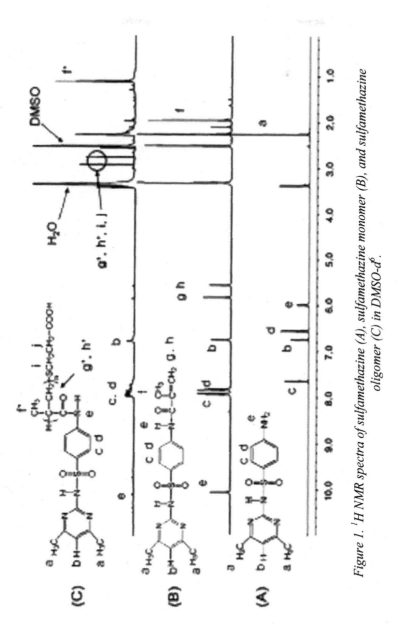

Figure 1. ^1H *NMR spectra of sulfamethazine (A), sulfamethazine monomer (B), and sulfamethazine oligomer (C) in DMSO-d^6.*

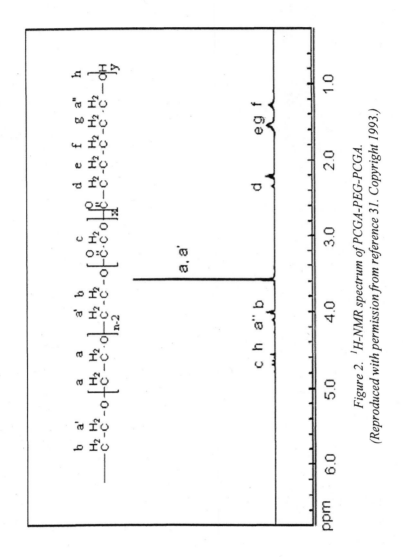

Figure 2. ^{1}*H-NMR spectrum of PCGA-PEG-PCGA.*
(Reproduced with permission from reference 31. Copyright 1993.)

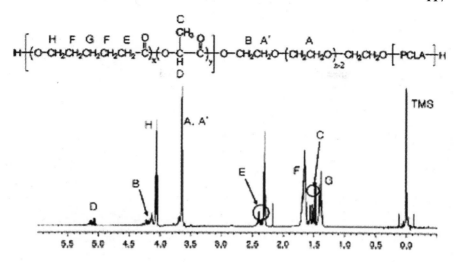

Figure 3. ^1H NMR spectrum of PCLA-PEG-PCLA block copolymer in CDCl$_3$ and its chemical structure.

labeled in Figure 3. Among the proton peaks, the methylene proton of the oxyethylene unit (A, A'), the methine proton of the LA unit (D), and the methylene proton (on the neighboring carbonyl group) of the CL unit (E) were used to calculate the Mn and composition of the block copolymer according to the method described in reference *(34)*. OSM was coupled with the temperature-sensitive block copolymers (PCLA-PEG-PCLA) or (PCGA-PEG-PCGA) using DCC and DMAP.

The synthesis of the OSM-PCLA-PEG-PCLA-OSM or OSM-PCGA-PEG-PCGA-OSM block copolymer was confirmed using ^1H NMR and GPC. The ^1H NMR spectrum of OSM-PCLA-PEG-PCLA-OSM and OSM-PCGA-PEG-PCGA-OSM show aromatic protons (7.6-8.0 ppm, c,d) and an imidazole ring proton (around 6.8 ppm, b), which are the typical signals associated with OSM (Figure 4). Also, the molecular weights of OSM-PCLA-PEG-PCLA-OSM and OSM-PCGA-PEG-PCGA-OSM showed a significant increase compared to those of PCLA-PEG- PCLA and PCGA-PEG-PCGA (Figure 2, 3). Tables II and III show the molecular weights of the triblock and penta-block copolymers. In addition, the GPC traces of OSM, PCLA-PEG-PCLA and OSM-PCLA-PEG-PCLA-OSM copolymer show a narrow molecular weight distribution (Figure 5). These results demonstrate that a block copolymer with a narrow molecular weight distribution was successfully synthesized.

118

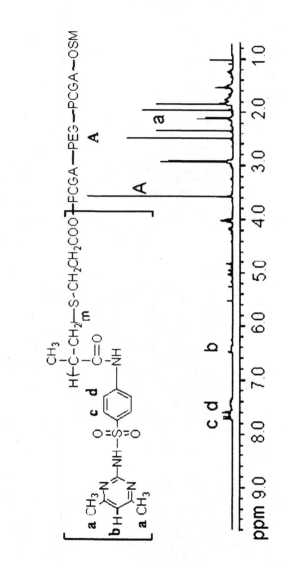

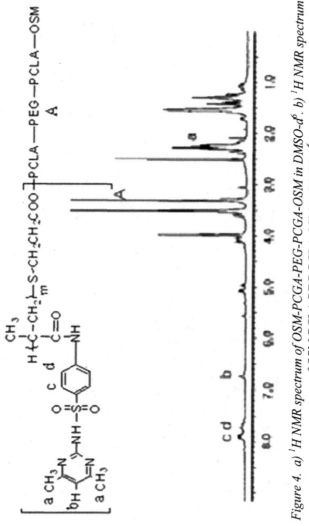

Figure 4. a) ^1H NMR spectrum of OSM-PCGA-PEG-PCGA-OSM in DMSO-d^6. b) ^1H NMR spectrum of OSM-PCLA-PEG-PCLA-OSM in DMSO-d^6.

Table II. Physical parameters of PCLA-PEG-PCLA and OSM-PCLA-PEG-PCLA-OSM block copolymers

OSM-PCLA-PEG-PCLA-OSM	PEG $M_n{}^b$	PEG/PCLA $(w/w)^a$	CL/LA mol/mol^a	OSM $M_n{}^c$	$M_w/M_n{}^c$
OSM$_1$-1384-1500-1384-OSM$_1$ (a-1-1)	1500	1/1.85	2.44/1	1144	1.45
OSM$_1$-1554-1500-1554-OSM$_1$ (a-2-1)	1500	1/2.08	2.59/1	1144	1.48
OSM$_2$-1554-1500-1554-OSM$_2$ (a-2-2)	1500	1/2.08	2.59/1	937	1.46
OSM$_1$-1642-1750-1642-OSM$_1$ (b-1-1)	1750	1/1.89	2.44/1	1144	1.50
OSM$_1$-1823-1750-1823-OSM$_1$ (b-2-1)	1750	1/2.08	2.49/1	1144	1.53
OSM$_1$-1856-2000-1856-OSM$_1$ (c-1-1)	2000	1/1.86	2.64/1	1144	1.54
OSM$_1$-2104-2000-2104-OSM$_1$ (c-1-1)	2000	1/2.10	2.71/1	1144	1.56

[a]PCLA-PEG-PCLA number-average molecular weights were calculated from ^{1}H-NMR.
[b]Provided by Aldrich. [c]Measured by GPC.

Table III. Physical parameters of PCLA-PEG-PCLA and OSM-PCLA-PEG-PCLA-OSM block copolymers

OSM-PCGA-PEG-PCGA-OSM	PEG $M_a{}^b$	PEG/PCGA $(w/w)^a$	CL/GA mol/mol^a	OSM $M_n{}^c$	$M_w/M_n{}^c$
OSM$_4$-1091-1000-1091-OSM$_4$ (A-1)	1000	1/2.18	2.34/1	806	1.30
OSM$_4$-1590-1500-1590-OSM$_4$ (B-1)	1500	1/2.12	2.35/1	806	1.34
OSM$_4$-2474-2000-2474-OSM$_4$ (D-1)	2000	1/2.47	2.37/1	806	1.50
OSM$_4$-1881-1750-1881-OSM$_4$ (C-2.1)	1750	1/2.15	2.32/1	806	1.35
OSM$_3$-1461-1750-1461-OSM$_3$ (C-1)	1750	1/1.67	2.26/1	904	1.35
OSM$_3$-1881-1750-1881-OSM$_3$ (C-2)	1750	1/2.15	2.32/1	904	1.35
OSM$_3$-2118-1750-2118-OSM$_3$ (C-3)	1750	1/2.42	2.26/1	904	1.36
OSM$_3$-2336-1750-1336-OSM$_3$ (C-4)	1750	1/2.67	2.29/1	904	1.36
OSM$_3$-2494-1750-2494-OSM$_3$ (C-5)	1750	1/2.85	2.28/1	904	1.35

[a]PCGA-PEG-PCGA number-average molecular weights were calculated from ^{1}H-NMR.
[b]Provided by Aldrich. [c]Measured by GPC.

SOURCE: Reproduced with permission from reference 31. Copyright 1993.

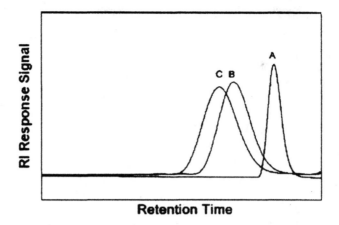

Figure 5. GPC traces. (A) PEG (Mn) 1500), (B) PCLA-PEG-PCLA (A-1),
(C) OSM-PCLA-PEG-PCLA-OSM (a-1-1).

Sol-Gel Phase Transition

The sol-gel transition phase diagrams of the block copolymer solutions were investigated under various pH and temperature conditions. Figure 6 shows the sol–gel transition mechanism of the penta-block copolymers. The copolymer solutions pass through 4 stages depending on the temperature and pH conditions. At low temperature (10°C) and high pH (8.0), PCLA or PCGA are not sufficiently hydrophobic and OSM is ionized and easy to dissolve. Therefore, the PCLA-OSM or PCGA-OSM block copolymers are slightly hydrophobic. For this reason, very small amounts of micelles form (stage D) and the copolymer solutions stay in the sol stage. When the temperature increases to 37°C (stage B), PCLA or PCGA become more hydrophobic, but OSM remains in the form of a hydrophilic block; if the pH decreases to 7.4 (stage C), OSM is de-ionized and becomes hydrophobic. Consequently, the hydrophobicity of PCLA-OSM or PCGA-OSM is increased and more micelles are formed, however the hydrophobicity of these blocks is not strong enough to form a hydrophobic link which would act as a bridge between the micelles, so the copolymer solutions still stay in the sol phase, although their viscosity increases. When the temperature is increased to 37°C and the pH reduced to 7.4 (stage A) both OSM and PCLA or PCGA become more hydrophobic, and the hydrophobicity of PCLA-OSM or PCGA-OSM is strong enough to form a lot of bridges between the micelles, with the result that the copolymer solutions stay in the gel phase.

122

a)

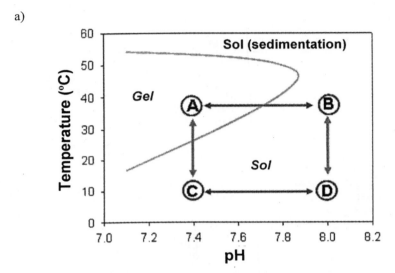

Figure 6. Sol-gel mechanism of the novel pH/temperature sensitive hydrogel.
a) Sol-gel transition phase diagram. b) The mechanism of sol-gel transition.
Continued on next page.

b)

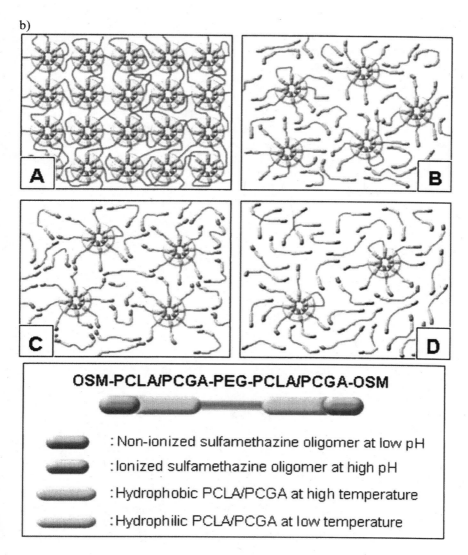

Figure 6. Continued.

Effect of Molecular Weight of OSM

Figure 7 shows the sol-gel diagrams of the OSM-PCLA-PEG-PCLA-OSM and OSM-PCGA-PEG-PCGA-OSM block copolymer solutions as a function of the molecular weight of the OSM component. In this figure, the critical gel pH (CGpH) and the lower (sol to gel transition) and upper (gel to sol transition) critical gel temperature (CGT) curves are shown. It is found that OSM, which is present mainly in an ionized state in the high-pH range, is present in the sol state regardless of its molecular weight. However, it can be seen that a decrease in pH results in the formation of a non-ionized OSM, which thus acts as a hydrophobic block. Also, particularly in the low-pH range, the hydrophobicity of the block copolymer increases with increasing molecular weight of OSM, resulting in an overall increase in the temperature range at which the gel forms. As a result, the CGpH in the phase diagram of C-2 is higher than that of C-2.1 (Figure 7a), and the CGpH in the phase diagram of a-2-1 is higher than that of a-2-2 (Figure 7b). In the case of almost every sample, when the temperature is increased from below the lower CGT to above it, the solution of the pH/temperature sensitive block copolymer transforms from the sol to the gel phase in a single stage. However, when the temperature is increased from below the upper CGT to above it, the gel-sol transition and suspension phase occur concurrently. When the temperature is greater than the upper CGT, the enthalpy of H_2O at this temperature is too high, so that the water is effectively liberated from the gel matrix.

Effect of PEG Molecular Weight

Figure 8 shows the changes in the sol-gel diagrams that occur as the molecular weights of PEG and the block copolymer increase, in the case where the molecular weight ratios of the hydrophilic (PEG) and hydrophobic (PCGA) or (PCLA) components are fixed. When the molecular weight of PEG was increased from 1000 to 2000, the lower CGT at a concentration of 10 wt% and pH 7.4 increased from 10°C (A-1) to 39°C (D-1), and the upper CGT increased from 38°C to 54°C (Figure 8a). Also, the lower CGT at a concentration of 15 wt% and pH 7.4 increased from 19°C (a-2-1) to 38°C (c-2-1), and the upper CGT increased from 45°C to 58°C (Figure 8b). It was found that the sol-gel phase diagram of the block copolymer moved toward higher temperatures with increasing block copolymer molecular weight at the same PEG/PCLA and PEG/PCGA ratio, yet revealed little or no change in the temperature range at which the block copolymer formed a gel. This suggests that when the length of the block copolymer is increased with a constant ratio of hydrophobic to hydrophilic blocks, gel formation by the block copolymer becomes possible, due to the more strongly hydrophobic conditions (i.e., strongly hydrophobic

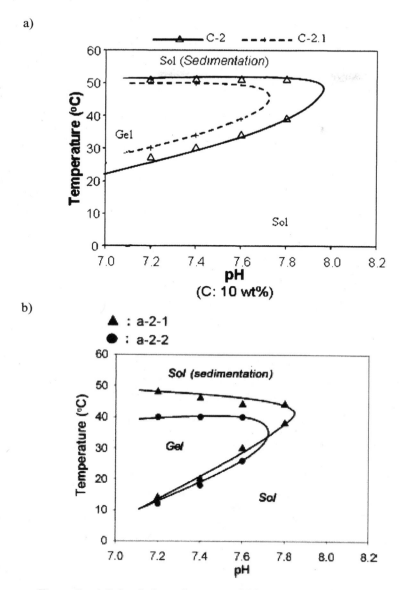

Figure 7. a) Sol-gel phase diagrams of OSM-PCGA-PEG-PCGA-OSM (C-2, C-2.1) block copolymer solutions with different molecular weight of sulfamethazine oligomers. Mn of PEG =1750; PEG/PCGA= 1/2.18 (w/w); concentration) 10%. b) Sol-gel phase diagrams of OSM-PCLA-PEG-PCLA-OSM (a-2-1, a-2-2) block copolymer solutions with different molecular weight of sulfamethazine oligomers. Mn of PEG =1500; PEG/PCLA= 1/2.08 (w/w); concentration) 15%.

a)

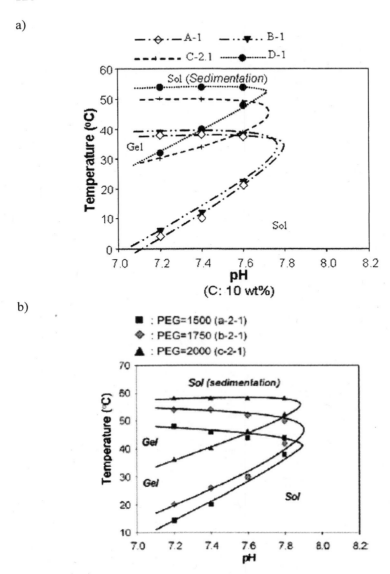

Figure 8. a) Sol-gel phase diagrams of OSM-PCGA-PEG-PCGA-OSM block copolymer solutions with different PEG molecular weights, similar PEG/PCGA ratios, and the same sulfamethazine oligomer. b) Sol-gel phase diagrams of OSM-PCLA-PEG-PCLA-OSM block copolymer solutions with different PEG molecular weights, similar PEG/PCLA ratios, and the same sulfamethazine oligomer. Concentration 15%. (a is reproduced with permission from reference 31. Copyright 1993.)

interactions at high temperature). It was also found that the temperature range at which the gels formed was affected mainly by the ratio of the hydrophobic to hydrophilic blocks. In addition, it was shown that, in the low pH range, the temperature range in which the gels formed decreased with increasing block copolymer molecular weight. This occurs because, regardless of the lengths of PEG and PCLA or PCGA in the various pH- and temperature-sensitive block copolymers, the OSM (with a constant molecular weight) is present in a non-ionized state at low pH and thus acts as a hydrophobic block, so that the ratio of the hydrophobic (PCLAOSM) or (PCGAOSM) to hydrophilic (PEG) blocks is decreased with increasing molecular weight of PEG. Accordingly, it was found that, at low pH, the temperature range in which a gel is formed is slightly decreased with increasing total molecular weight of the block copolymer.

Effect of Hydrophobic/Hydrophilic Ratio

Figures 9 and 10 shows the phase diagrams of these penta-block copolymers with various hydrophobic/hydrophilic chain lengths. As the hydro-phobic/hydrophilic chain length ratio is increased from 1.67 (C-1) to 2.85 (C-5) (Figure 9) and from 1.85 (a-1-1) to 2.08 (a-2-1) (Figure 10), the critical gel temperature (CGC) of the penta-block copolymers decreases in a proportional manner, the CGpH increases, and the range between the lower CGT and upper CGT increases. As the PCGA/PEG or PCLA/PEG ratio increases, the hydrophobic chain length becomes longer, and so the hydrophobicity of the copolymer becomes greater, so that the resulting micelles are generally formed with fewer associated chains *(34)*. As more bridging connections are formed between the micelles, more grouped micelles are produced, and hence the sol to gel and gel to sol phase transitions occur at lower and higher CGTs, respectively. As a result, the gel phase areas in the phase diagrams of these penta-block copolymers could be controlled by changing the hydrophobic/hydrophilic length ratios. Thus, the gel phase zone is dependent on the hydrophobic change length of the pH/temperature sensitive copolymers.

Concentration Dependence

Figure 11 shows the phase diagrams of the OSM-PCLA-PEG-PCLA-OSM block copolymer solution resulting from the changes in pH and temperature at various concentrations. At a concentration of more than 10 wt %, the block copolymer solution formed a gel as the temperature increased in the low-pH region (below pH 7.4). In this case, as the concentration was increased, the sol-to-gel transition temperature decreased and the gel region became wider. At low concentrations, the gelling of the block copolymer solution became possible, due

a)

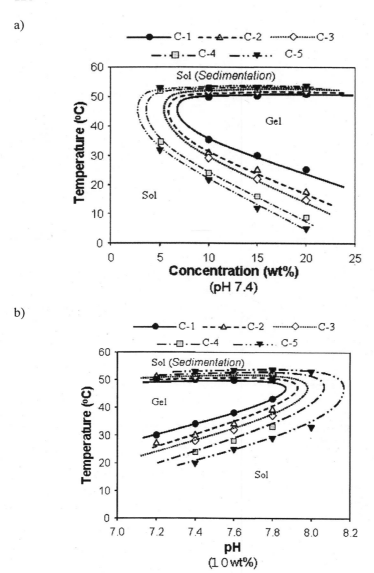

b)

Figure 9. a) Sol-gel transition phase diagrams of OSM-PCGA-PEG-PCGA-OSM with various hydrophobic/hydrophilic chain lengths and concentrations at pH 7.4. b) Sol-gel transition phase diagrams of OSM-PCGA-PEG-PCGA-OSM with various hydrophobic/hydrophilic chain lengths and pH values at concentration 10 wt%. (Reproduced with permission from reference 31. Copyright 1993.)

a)

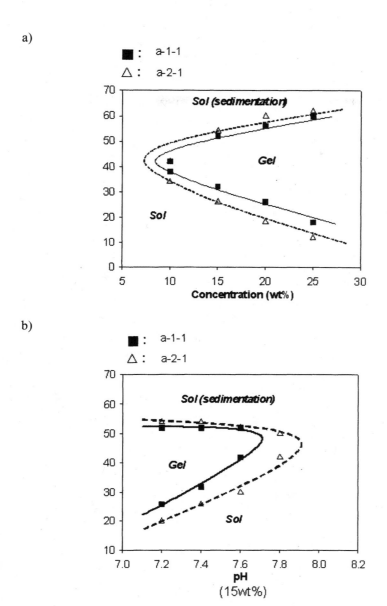

b)

Figure 10. a) Sol-gel transition phase diagrams of OSM-PCLA-PEG-PCLA-OSM with various hydrophobic/hydrophilic chain lengths and concentrations at pH 7.4. b) Sol-gel transition phase diagrams of OSM-PCLA-PEG-PCLA-OSM with various hydrophobic/hydrophilic chain lengths and pH values at concentration 15 wt%.

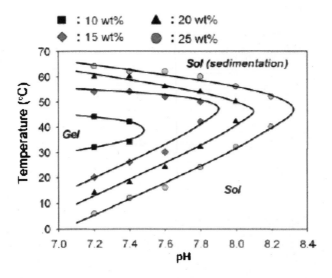

Figure 11. Sol-gel phase diagrams of OSM-PCLA-PEG-PCLA-OSM (b-2-1) block copolymer solutions at various concentrations.

to the more strongly hydrophobic conditions, i.e., the more strongly hydrophobic PCLA was present at high temperatures and the mostly non-ionized OSM was present at low pH. On the other hand, at high concentration, the gelling of the block copolymer solution was found to occur easily under more weakly hydrophobic conditions, although it was difficult to solubilize the block copolymer at a concentration of more than 25 wt% in the buffer solution. At high concentrations (above 25 wt%), the block copolymer solution formed gels even at 0°C and showed no evidence of a sol state at low temperatures in the pH range used in these experiments (pH 7.2-8.0).

The typical phase diagrams exhibited a CGC, CGpH and two CGT curves. The CGC and CGpH of the pH/temperature sensitive block copolymers could be controlled by varying the ratio of the hydrophobic/hydrophilic chain lengths, the concentration of the copolymer in the solution, and the molecular weight of the pH sensitive component. In other words, the gel phase zones of these penta-block copolymers could be adjusted by varying the molecular weight of PEG, the PCGA/PEG or PCLA/PEG ratio, the OSM length and the concentration of the copolymers.

Degradation of these Block Copolymers

Figure 12 shows the degradation behaviors of the PCGA-PEG-PCGA and OSM-PCGA-PEG-PCGA-OSM block copolymer solutions compared to those of PCLA-PEG-PCLA and OSM-PCLA-PEG-PCLA-OSM at 37°C and pH 7.4 for a

PCLA-PEG-PCLA and OSM-PCLA-PEG-PCLA-OSM at 37°C and pH 7.4 for a specified period. As can be seen in Figure 12, the degradation slopes for PCGA-PEG-PCGA, PCLA-PEG-PCLA and OSM-PCLA-PEG-PCLA-OSM are similar. In the case of OSM-PCGA-PEG-PCGA-OSM, the rate of degradation is particularly high in the early stages, as represented by the steep gradient. After 18 days, however, the rate of degradation in aqueous solution is observed to decrease. Interestingly, the molecular weight of these polymers was also observed to decrease after 40 days of degradation. PCGA-PEG-PCGA and OSM-PCGA-PEG-PCGA-OSM demonstrated losses of 101 and 2323, respectively, compared to PCLA-PEG-PCLA and OSM-PCGA-PEG-PCGA-OSM, which showed losses of 934 and 998, respectively (calculated from Figure 12). These results indicate that OSM-PCGA-PEG-PCGA-OSM degrades faster than OSM-PCGA-PEG-PCGA-OSM. As such, it is expected that drugs will be released faster when loaded in OSM-PCGA-PEG-PCGA-OSM as compared to OSM-PCLA-PEG-PCLA-OSM.

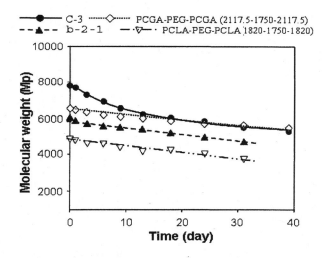

Figure 12. Degradation of block copolymers. (Reproduced with permission from reference 31. Copyright 1993.)

Sol-Gel by Injection into Environment (In Vitro Test)

As can be seen in Figure 8b and 9a the physiological conditions (37°C and pH 7.4) are located in the center of the gel zone on the sol-gel phase diagrams of samples b-2-1 and C-5. Therefore, these samples were used for the sol–gel transition in vitro test. The results showed that these copolymer solutions can form a gel after being injected into PBS buffer under physiological conditions (Figure 13).

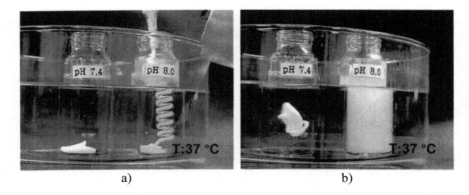

a)   b)

Figure 13. Sol-gel injection of penta-block (b-2-1) by in vitro test.
a) Injection to environment. b) After injection and shaking.

Cytotoxicity of Hydrogel

Figure 14 shows the cytotoxicity of OSM-PCLA-PEG-PCLA-OSM (b-2-1). The results show that these copolymers are non toxic and could be used as biomaterials.

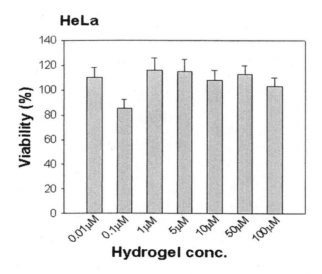

Figure 14. Cytotoxicity of OSM-PCLA-PEG-PCLA-OSM (b-2-1).

Sol-Gel Transition by In Vitro Test

A 20 wt% solution of the copolymer at 10°C and pH 8.0 was (prepared?) from OSM-PCLA-PEG-PCLA-OSM (b-2-1) to check the sol-gel transition by the in vivo test. The copolymer solution was injected into a rat (model male SD), and the rat was operated on to check whether a gel was formed 10 min after the injection. The results in Figure 15 show that the copolymer solution can form a gel in the rat body 10 min after being injected.

Drug Loading and Release

Figure 16 shows the results for the release into the environment of the drug paclitaxel (PTX), which was loaded into the OSM-PCGA-PEG-PCGA-OSM

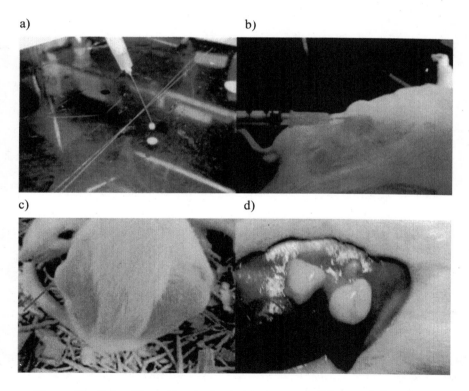

Figure 15. Sol-gel transition by in vivo test on male SD rat. a) Copolymer solution. b) Injecting to male SD rat. c) Injected sites on rat. d) Gel was formed after 10 min injected.

134

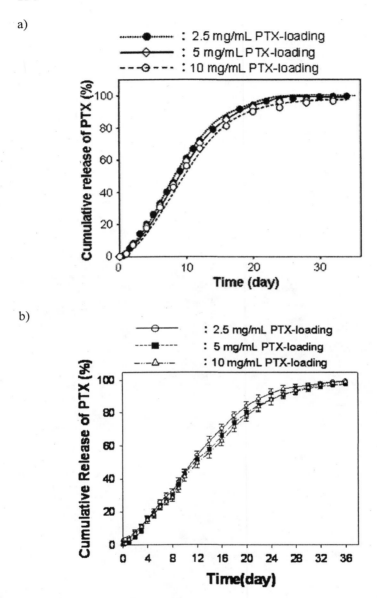

Figure 16. PTX releasing by in vitro test. a) PTX which was loaded into OSM-PCGA-PEG-PCGA-OSM (C-3) is releasing. b) PTX which was loaded into OSM-PCLA-PEG-PCLA-OSM (b-1-1) is releasing. (Reproduced with permission from reference 31. Copyright 1993.)

(C-3) (Figure 16a) and OSM-PCLA-PEG-PCLA-OSM (b-1-1) (Figure 16b) matrices, obtained by the in vitro test method (the drug content in the polymer aqueous solution was measured in weight percent compared to the matrix). PTX is a hydrophobic drug; it is loaded into the hydrophobic core of the hydrogel easily. The release of the drug occurs after the disintegration of the hydrophobic core of the hydrogel. As the result, the release of PTX from the gel is dependent on the degradation of the penta-block copolymer. As shown in Figure 12, the degradation of OSM-PCGA-PEG-PCGA-OSM is faster than that of OSM-PCLA-PEG-PCLA-OSM, so the release of PTX from OSM-PCGA-PEG-PCGA-OSM is faster than that from OSM-PCLA-PEG-PCLA-OSM. Figure 16. a shows that more than 90% of the PTX was released from the OSM-PCGA-PEG-PCGA-OSM matrix after 20 days, while the same amount of PTX was released from the OSM-PCLA-PEG-PCLA-OSM matrix after 30 days (Figure 16b). When a higher concentration of drug was loaded into the matrix, the amount of drug remaining in the matrix after its release was found to be higher.

Summary

In this study, pH- and temperature-sensitive block copolymers, with varying PEG lengths, PEG/PCLA and PEG/PCGA ratios, and molecular weights of OSM, were synthesized through a combination of ring-opening reactions and DCC coupling. Their aqueous solutions show a sensitive, reversible sol-to-gel transition over a small pH range (pH 7.2-8.0) and temperature range (10-50°C), which are representative of the conditions found in the body. The sol-gel phase diagrams of the block copolymer solutions were deliberately controlled by altering the ratio of hydrophobic to hydrophilic blocks within the block copolymer, the PEG length, or the molecular weight of the OSM component. Accordingly, it could be confirmed that the temperature and pH ranges at which the sol-gel transition occurs can be adjusted by varying the molecular weight and composition ratio of the block copolymer. The degradation of these copolymers could be controlled by using different hydrophobic blocks (PCLA or PCGA) to form the penta-block copolymers. The aqueous solutions of the block copolymer with the compositions : PEG =1750, PEG/PCLA ratio = 1/2.08 and OSM = 1144 and PEG =1750, PEG/PCG A ratio from 1/2.42 to 1/2.85 and OSM = 904 formed the most stable gels in simulated in vivo conditions (37°C, pH 7.4). The results of the sol-gel transition experiments in vitro and in vivo showed that at pH 8.0, the viscosity of the block copolymer solutions increased only very slightly with increasing temperature, with no obvious gel formation, implying that this pH- and temperature-sensitive block copolymer solution can be easily injected into the body. Moreover, the injected block copolymer solution is able to form a gel as a result of a small pH change from pH 8.0 to pH 7.4 in vivo. In addition, these copolymers are non toxic, so these hydrogels have the potential to

136

be used as an injectable carrier for sustained release drug delivery systems. The PTX loading and release in vitro test indicated that the release of the drug could be controlled by adjusting the degradation of the hydrogel copolymer.

References

1. Jeong, B.M.; Kim, S. W.; Bae, Y. H. *Adv. Drug Delivery Rev.* **2002**, *54*, 37.
2. Hoffman, A. S. *Adv. Drug Delivery Rev.* **2002**, *43*, 3.
3. Takashi, M.; Tadashi, U.; Katsuhiko, N. *Adv. Drug Delivery Rev.* **2002**, *54*, 79.
4. Peppas, N. A. *Hydrogels in Medicine and Pharmacy;* CRC Press, Boca Raton FL, 1987.
5. Ding, Z.; Fong, R. B.; Long, C. J.; Stayton, P. S.; Hoffman, A. S. *Nature* **2001**, *411*, 59.
6. Yoshida, R.; Sakai, K.; Okano, T.; Sakurai, Y. *J. Biomater. Sci.Polym. Ed.* **1994**, *6*, 585.
7. Hassan, C. M.; Doyle, F. J., III.; Peppas, N. A. *Macromolecules* **1997**, *30*, 6166.
8. Temenoff, J. S.; Mikos, A. G. *Biomaterials* **2000**, *21*, 2405.
9. Tanaka, T.; Fillmore, D.; Sun, S.T.; Nishio, I.; Swislow, G.; Shah, A. *Phys. Rev. Lett.* **1980**, *45*, 1636.
10. Hirokawa, Y.; Tanaka, T. *J. Chem. Phys.* **1984**, *81*, 6379.
11. Yoshida, R.; Uchida, K.; Taneko, T.; Sakai, K.; Kikuchi, A.; Sakurai, Y.; Okano, T. *Nature* **1995**, *374*, 240.
12. Tanaka, T.; Nishio, I.; Sun, S. T.; Nishio, S. U. *Science* **1982**, *218*, 467.
13. Osada, Y.; Okuzaki, H.; Hori, H. *Nature* **1992**, *355*, 345.
14. Irie, M. *Adv. Polym. Sci.* **1993**, *110*, 49.
15. Suzuki, A,; Tanaka, T. *Nature* **1990**, *346*, 345.
16. Yang, Z.; *Pickard*, S.; Deng, N. J.; Barlow, R. J.; Attwood, D.; Booth, C. *Macromolecules* **1994**, *27*, 2371.
17. Malmsten, M.; Lindman, B. *Macromolecules* **1992**, *25*, 5440.
18. Hatefi, A.; Amsden, B. *J. Controlled Release* **2002**, *80*, 9.
19. Malmsten, M.; Lindman, B. *Macromolecules* **1992**, *25*, 5446.
20. (a) Yang, Z.; Pickard, S.; Deng, N.; Barlow, R. J.; Attwood, D.; Booth, C. *Macromolecules* **1994**, *27*, 2371. (b) Alexandridis, P.; Hatton, T. A. *Colloids Surf. A: Physicochem. Eng. Aspects* **1995**, *96*, 1.
21. (a) Rashkov, I.; Manolova, N.; Li, S. M.; Espartero, J. L.; Vert, M. *Macromolecules* **1992**, *29*, 50. (b) Cerrai, P.; Tricoli, M.; Lelli, L.; Guerra, G. D.; Sbarbati Del Guerra, R.; Cascone, M. G.; Giusti, P. *J. Mater. Sci., Mater. Med.* **1994**, *5*, 308.
22. (a) Chenite, A.; Chaput, C.; Wang, D.; Combes, C.; Buschmann, M. D.; Hoemann, C. D.; Leroux, J. C.; Atkinson, B. L.; Binette, F.; Selmani, A.

Biomaterials **2000**, *21*, 2155. (b) Han, H. D.; Nam, D. E.; Seo, D. H.; Kim, T. W.; Shin, B. C. *Macromolecular Res.* **2004**, *12*, 507.

23. (a) Tanigami, T.; Ono, T.; Suda, N.; Sakamaki, Y.; Yamaura, K.; Matsuzawa, S. *Macromolecules* **1989**, *22*, 1397. (b) Lee, B. H.; Lee, Y. M.; Sohn, Y. S.; Song, S. C. *Polym. Int.* **2002**, *51*, 658.

24. Liang, H.; Hon, M.; Ho, R.; Chung, C.; Lin, Y.; Chen, C.; Sung, H. *Biomacromoleules* **2004**, *5*, 1917.

25. (a) Lee, D. S.; Shim, M. S.; Kim, S. W.; Lee, H.; Park, I.; Chang, T. *Macromol. Rap. Commun.* **2001**, *22*, 587. (b) Jeong, B. M.; Bae, Y. H.; Kim, S. W. *J. Controlled Release* **2000**, *63*, 155.

26. (a) Middaugh, C. R.; Evans, R. K.; Montgomery, D. L.; Casimiro, D. R. *J. Pharm. Sci.* **1998**, *87*, 130. (b) Fu, K.; Pack, D. W.; Klibanov, A. M.; Langer, R. *Pharm. Res.* **2000**, *17*, 100.

27. Kim, B. S.; Flamme, K. L.; Nicolas, A. P. *J. Appli. Polym. Sci.* **2003**, *89*, 1606.

28. Kim, B. S.; Nicolas, A. P. *J. Biomater. Sci. Polymer Edn.* **2002**, *13*(11), 1271.

29. Shim, W. S.; Yoo, J. S.; Bae, Y. H.; Lee, D. S. *Biomacromolecules* **2005**, *6*, 2930.

30. Shim, W. S.; Kim, S. W.; Lee, D. S. *Biomacromolecules* **2006**, *7*(6), 1935.

31. Stevens, M. G.; Olsen, S. *J. Immunol Methods* **1993**, *157*, 225.

32. Luo, D.; Saltzman, W. M. *Nat. Biotechnol.* **2000**, *18*, 33.

33. Jeong, B. M.; Lee, D. S.; Shon, J.; Bae, Y. H.; Kim, S.W. *J. Polym. Sci. Part A* **1999**, *37*, 751.

34. Jeong, B. M.; Bae, Y. H.; Kim, S.W. *Macromolecules* **1999**, *32*, 7064.

Chapter 8

Engineering of Solid Lipid Nanoparticles for Biomedical Applications

Eric M. Sussman, Ashwath Jayagopal, Frederick R. Haselton, and V. Prasad Shastri

Department of Biomedical Engineering, Vanderbilt University
5824 Stevenson Center, Nashville, TN 37235
Corresponding author: prasad.shastri@vanderbilt.edu

Solid Lipid Nanoparticles (SLN) bear significant potential as drug delivery systems and diagnostic probes. They constitute a promising alternative to polymeric nanoparticle and liposomal formulations due to their efficient loading and stabilization of poorly soluble or unstable compounds, and the potential for scaling up the production processes for pharmaceutical applications. To enhance the scope of their biomedical application, methods which enable simultaneous drug/probe entrapment and presentation of surface functionalities are desired. We have developed a process for the single-step formulation of functionalized polymeric nanoparticles with tunable size and narrow polydispersity (*1*). By incorporation of a polyelectrolyte or polymer within the aqueous phase, our process enables the surface engineering of diverse surface moieties upon a non-functionalized polymer backbone within a binary solvent system, via a phase inversion process. By substitution of a biocompatible coco-glyceride for the non-functionalized polymer in the organic phase, we have applied this strategy toward the formulation of surface functionalized SLN capable of entrapping physicochemically-diverse cargoes. We investigated the effect of process parameters on nanoparticle size, surface charge, and compound entrapment efficiency. Species to be encapsulated included fluorescent dyes, semiconducting nanocrystals, and superparamagnetic iron oxide nanoparticles (SPIO). Surface moieties incorporated in the aqueous phase included PEG, amine groups and carboxyl groups for the bioconjugation of targeting ligands. Polymeric NP and SLN were characterized by laser light scattering, x-ray photoelectron spectroscopy (XPS), fluorescence spectrophotometry, electron microscopy, and fluorescence microscopy. Our studies demonstrate that the solvent composition can be varied to control nanoparticle hydrodynamic radius while maintaining a narrow polydispersity. Both hydrophilic and hydrophobic species can be efficiently entrapped within the lipid matrix to create targeted and stable high signal intensity diagnostic probes or concentrated drug carriers for biomedical applications. We anticipate the application of SLN toward magnetic resonance and fluorescence imaging, targeted chemotherapy, as well as the surface modification of biomaterials for tissue engineering strategies.

Nanoparticulate delivery systems generally aim to safely and efficiently deliver therapeutic or diagnostic compounds to specific tissue sites, without degradation of the cargo that would limit its efficacy. Nanoparticles with solid lipid matrices, or solid lipid nanoparticles, (SLN) exhibit several key features which make them promising agents for the delivery of diagnostic probes or therapies *in vivo*. Among these are their capacity to entrap poorly water-soluble compounds, stability (remaining solid at physiologic temperatures and/or during storage), and biocompatibility (2-4). They are not limited by some of the challenges associated with current nanoparticulate drug delivery systems in research and clinical administration stages, including poly-lactide-*co*-glycolide nanoparticles (PLGA NP) and liposomes, which have been associated with acidic degradation of protein cargoes (5) and potentially low cargo loading and stability (6), respectively.

In order to expand the capabilities of SLN in medical imaging and drug delivery and to facilitate their clinical implementation, further work is warranted in the continued adaptation of laboratory-scale production processes for the pharmaceutical-scale, and the modifications of process parameters to control efficiency-critical properties of SLN, such as size and loading capacity. Furthermore, it is important that SLN be capable of bearing surface functionalities and/or functionally-diverse cargoes such as contrast agents and drugs, regardless of their physicochemical properties (e.g. hydrophilicity, size).

Such a versatile carrier technology at the nanometer scale of biology would provide a template for the manufacture of a broad spectrum of drugs and imaging agents which work at the scale of biology.

We have developed a single-step process for the rapid synthesis of functionalized polymeric nanoparticles based on the polymers PLGA and poly-lactic acid (PLA) (*1*). The solvation core of this non-functionalized polymer backbone is specifically engineered such that addition of a polyelectrolyte or polymer-containing aqueous phase results in the instantaneous formation of a nanoparticle, the surface of which bears the aqueous moiety. We have adapted this process toward the surface engineering of SLN, enabling them to load various types of cargoes while at the same time bearing different surface functionalities. Our process obviates the need for surfactants, emulsifiers, and melting/cooling steps which are often involved in other SLN synthesis methods. Tailoring of the solvent composition and presence of aqueous phase moieties in the process was found to have profound effects on the surface engineering and encapsulation features of the nanoparticulate system. We discuss the synthesis and characterization of polymeric and solid lipid NP and implications for biomedical applications such as drug delivery, imaging, and tissue engineering.

Results and Discussion

Control of Polymeric Nanoparticle Size and Polydispersity

We hypothesized that judicious selection of an appropriate organic solvent composition for dissolution of the non-functionalized polymer (i.e., PLGA or PLA), by alteration of solvent polarity, would have significant effect on NP size by modulating the rate of polymer chain collapse into a nanoparticle upon the rapid addition of water. By varying both the choice of solvent pairs and the relative solvent volumetric ratios, the process yielded polymeric NP diameters ranging from 70-400nm, with narrow PDI. Representative results are shown in Table I. It is likely that the primary determinant of NP size in our system is solvent composition. Specifically, a highly-polar solvent system with dissolved PLGA or PLA, upon addition of water, collapses rapidly into small NP, whereas a less polar solvent system would support slower polymer chain collapse upon interaction with water, to yield larger NP.

Surface Engineering of Polymeric Nanoparticles

The presence of a polyelectrolyte in the aqueous phase prior to organic-aqueous phase mixture provided a mechanism for spontaneous surface functionalization of the solidifying non-functionalized polymer core. We

142

Table I. Typical NP Diameter and Polydispersity Index (PDI) Measured for Different Combinations of Polymer, Polyelectrolyte (Aqueous), and Binary Solvent Systems

Polymer	Polyelectrolyte	Solvent System	Size (nm)	PDI
PLGA	None	THF/Acetone	243	0.05
PLGA	PAA	THF/Acetone	271	0.07
PLGA	PSS	THF/Acetone	404	0.09
PLGA	PLL	THF/Acetone	259	0.07
PLA	None	Dichloromethane/Acetone	184	0.08

SOURCE: Reproduced with permission from reference 1. Copyright 2004 Materials Research Society.

hypothesized that by utilizing polyelectrolytes for surface functionalization with high affinities for water, preferential accumulation of the surface moieties would occur at the NP-water interface following phase inversion. Zeta potential (ζ) analysis of polymeric NP supports our hypothesis. Figure 1 shows the surface charge variation as a function of pH for non-functionalized (PLGA) and functionalized (PLGA-PSS, PLGA-PLL) nanoparticles. As shown, the isoelectric point of the nanoparticulate formulation is associated closely with the pK_a of the ionizable group (PLGA = -COOH, PSS = SO_3H, PLL = NH_2).

XPS analysis also confirms our hypothesis, as polymeric NP surface compositions were abundant with the desired functionality (Table II). It is important to note that PSS, despite its charge similarity to PLGA (negative charge), is effectively entrapped on the PLGA NP surface using our approach (66%). Additionally, the ability to functionalize NP surfaces with PLL enables the potential bioconjugation of ligands via amide linkages to the NP surface for *in vivo* targeting applications. Thus we are able to rapidly produce PLGA NP of defined size ranges and narrow PDI, with the added benefit of customizing NP surface information.

Encapsulation of Small Molecules by Polymeric NP

Our single-step process for the surface engineering of polymeric NP does not appear to preclude the encapsulation of compounds for drug delivery or imaging purposes. Fluorescence microscopy of DAPI-loaded PLGA-PSS NP reveals that the water-soluble dye is efficiently and homogenously entrapped within the NP core (Figure 2). In other work, we have entrapped other fluorescent dyes, proteins, and synthetic drugs within polymeric NP. Thus, our process appears to be a versatile approach for packaging structurally and functionally-diverse cargoes within well-defined (tight size range and PDI) and surface functionalized NP.

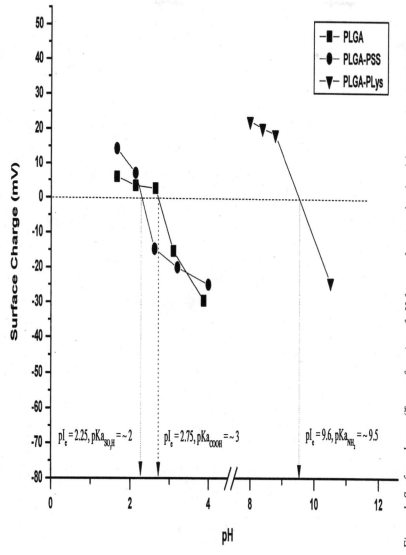

Figure 1. Surface charge (ζ) as a function of pH for non-functionalized and functionalized PLGA NP. Isoelectric points of NP correlate with the pKa of the PLGA or polyelectrolyte functional groups. (Reproduced with permission from reference 1. Copyright 2004 Materials Research Society.)

144

Table II. Surface Composition of PLGA NP as Analyzed by XPS (C^1S).

	COOR	COO-C-OR	C-OR	C-H_x	% of Surface Mass for Functional Moiety
PLGA	38.2±0.0	36.7±0.0	--	25.1±0.0	N/A
PLGA-PLL	27.1±3.1	28.0±3.2	9.0±1.6	36.0±4.7	24±9%
PLGA-PSS	12.1±1.6	12.5±1.6	--	75.5±3.2	66±4%

SOURCE: Reproduced with permission from reference 1. Copyright 2004 Materials Research Society.

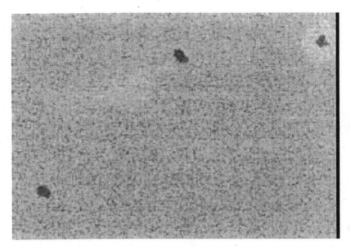

Figure 2. Fluorescence micrograph (400X) of DAPI-loaded PLGA NP. (Reproduced with permission from reference 1. Copyright 2004 Materials Research Society.)

Control of Solid Lipid Nanoparticle Size and Loading Efficiency

An adaptation of our procedure for synthesizing functionalized polymeric NP for the entrapment and targeted delivery of small therapeutic and diagnostic agents enabled us to develop a similar process for the rapid single-step production of functionalized SLN. After considerable investigation we found the binary solvent system NMP:Acetone to provide effective size control of SLN while also providing for excellent solubility of the coco-glyceride. High ratios of NMP:Acetone are associated with small SLN. Lower ratios result in larger SLN, likely due to inefficient lipid packing upon exposure to water. Quantum dots, which without commercial modifications are highly hydrophobic, were selected as a candidate for encapsulation, as it has a high molecular weight and has emerged as a fluorescent probe with superior optical properties for bioimaging applications (7). By varying the volumetric ratio of NMP:Acetone from 40:60 to 80:20, SLN bearing quantum dots ranging from 75nm-790nm were synthesized. Furthermore, by varying the SLN diameter through solvent composition, we were able to concurrently modulate SLN loading efficiency of QD (Figure 3). Larger SLN diameters, or larger lipid matrix cores, allow for more incorporation of QD as they partition into the lipid phase upon exposure to water. The high loading capacity of 100nm SLN, for example (100-150 QD avg. per SLN (n=10) as determined by TEM), allows for the creation of nanoparticulate probes with significantly enhanced signal intensity compared to single nanoparticles (Figure 3). This feature could be useful in many biomedical applications, including the magnetic resonance (e.g., CLIO-NP) or fluorescence-based (e.g., QD) signal amplification of low levels of antigens on cell surfaces, or the delivery of large payloads of drug into cells.

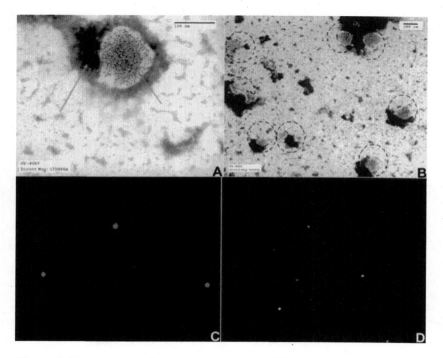

Figure 3. Transmission electron microscopy and fluorescence microscopy of SLN-QD. First row: A.) Transmission electron micrograph (175000X at 80keV, scale bar 100nm) showing 2.4nm 580nm-emitting CdSe/ZnS quantum dots (red arrow) within a lipid matrix (white background of QD) against phosphotungstic acid negative lipid stain (green arrow). B.) TEM (66000X at 80 keV) of same SLN-QD sample, showing dispersion and narrow size range of SLN-QD (outlined in blue). Second row: Fluorescence micrographs (40X) of large-diameter (C) and small-diameter (D) SLN-QD.

Surface Engineering of SLN

By incorporation of polyelectrolytes into the aqueous phase prior to the lipid-packing phase inversion step, we efficiently loaded the desired surface functionality upon the SLN surface in a manner analogous to that seen for polymeric NP. Functionalization of negative charges upon SLN (SLN-PSS) was confirmed by ξ analysis (Figure 4). As shown in Figure 5, fluorescence microscopy and spectrophotometry were indicative of efficient surface incorporation of dye-labeled PLL. PLL was readily functionalized upon SLN surfaces by layer-by-layer assembly. By first synthesizing SLN initially coated with the negatively-charged PSS, which effectively incorporates within the lipid while remaining preferentially at the NP-water interface (due to the sulfonic acid group), the positively-charged PLL can be layered electrostatically. Other surface functionalities presented on SLN surfaces to date include polyethylene glycol (PEG) for evasion of the mononuclear phagocyte system *in vivo* (*8*), polyacrylic acid (PAA), which has been utilized to incorporate mucoadhesive properties (*9*), and streptavidin for biotin-based immunofluorescence applications (*10*). Thus, our process enables the encapsulation of functionally-diverse nanoparticles (i.e. dye, contrast agent, therapeutic) within solid lipid matrices capable of bearing different types of surface information.

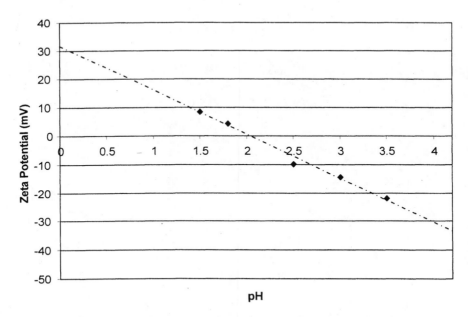

Figure 4. Zeta potential analysis of SLN-QD-PSS as a function of pH. The isoelectric point of the sample correlates with the pKa of the sulfonic acid group of PSS, thus indicating effective surface functionalization.

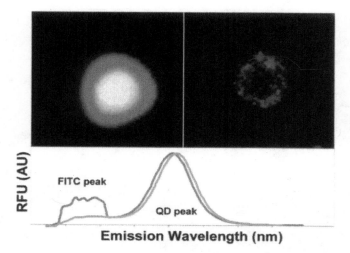

Figure 5. Top Row: Fluorescence microscopy of QD entrapped within PSS-FITC/PLL-functionalized SLN, in the QD 580 emission (left) and FITC emission channels (right). The core of the SLN exhibits high QD fluorescence due to effective entrapment of hydrophobic QD within the lipid matrix, whereas FITC-PLL is adsorbed upon previously incorporated PSS at the NP-water interface, thus providing an amine template for further bioconjugation. Bottom Row: Fluorescence spectrophotometry of QD-PSS (orange) and QD-PSS-FITC/PLL (green) SLN. Surface functionalization of PLL (initial FITC peak as indicated) does not significantly affect native QD fluorescence (580nm emission peak, as indicated).

Multimodal SLN for Biomedical Applications

In order to demonstrate the potential of SLN as multimodal carriers, we co-encapsulated 50nm silica-capped SPIO and FITC-BSA within PSS-functionalized lipid matrices by incorporation of both species within the aqueous phase prior to phase inversion. FITC-BSA/SPIO-loaded SLN coated with PSS were incubated in a solution containing 0.5% trypan blue and analyzed by fluorescence microscopy (Figure 6). Trypan Blue is an effective quenching agent for reducing extracellular FITC as part of phagocytosis assays (*11*). We therefore utilized trypan blue to quench unencapsulated FITC-BSA to confirm that only FITC-BSA within SLN was visualized. Upon application of a magnetic field using a static magnet (1.5 T), FITC-BSA nanoparticles shown in Figure 6 were observed to move in response to magnet polar orientation. Fluorescent nanoparticles were highly sensitive to rapid changes in the external field while control nanoparticles consisting of FITC-BSA mixed with SPIO in distilled water without lipid did not move in response to changes in magnetic

Figure 6. Fluorescence microscopy of SLN bearing FITC-BSA and SPIO in a solution of 0.5% Trypan Blue. Encapsulated FITC-BSA not accessible to Trypan Blue in mounting medium is evident, and all SLN were observed to move toward the pole of an externally applied magnet, indicating successful co-encapsulation of FITC-BSA with SPIO.

fields. These data suggest that FITC-BSA was co-encapsulated with SPIO, and that the native magnetization properties of the NP are not affected by encapsulation. We have in effect created a technique applicable to the development of magnetooptical probes which employ fluorophores (e.g. quantum dots or fluorescently-labeled proteins) and SPIO, which would have applications for in vivo imaging using MRI and fluorescence imaging techniques such as multiphoton excitation microscopy. Further work is directed toward the co-encapsulation of QD and SPIO to develop high-intensity magnetooptical probes for bioimaging applications, as well as drug-QD conjugates for pharmacokinetic/pharmacodynamic studies of chemotherapeutics.

Conclusion

We have developed a technique for the rapid synthesis of functionalized polymeric NP, and an analogous procedure for producing functionalized solid lipid nanoparticles. Our process provides for the simultaneous, efficient entrapment of compounds of varying physicochemical properties, so that combinative imaging and therapeutic strategies can be harnessed. Mechanical dispersion, the addition of emulsifying agents, or temperature-dependent steps are not required. The lipid utilized in SLN is commonly used in cosmetic formulations, and all solvents used in synthesis are removed, thus at least ensuring percutaneous biocompatibility. However, this process should be readily applicable to the formulation of SLN using lipids known to possess satisfactory *in vivo* biocompatibility. The process can be easily scaled-up for the mass production of pharmaceuticals and SLN are highly-stable in storage. Future work is directed toward with the applications of multimodal SLN for the simultaneous utilization of multiple imaging modalities as well as combined drug delivery/gene therapy approaches.

Materials and Methods

Materials

Poly(dl-lactide-*co*-glycolide) (PLGA, RG 503, MW=30,000) and poly (L-lactide) (PLA) MW = 70,000, inherent viscosity 1.20 dL/g in CHCl₃) were purchased from Birmingham Polymers. Prior to use, PLGA and PLA were purified by precipitation from methylene chloride (MeCl) in methanol. Tetrahydrofuran (THF), Toluene, MeCl, Acetone, and 1-Methyl-2-pyrrolidinone (NMP) were purchased from Sigma or Fisher Scientific (HPLC grades). HCl, NaOH, Fluorescein isothiocyanate-bovine serum albumin conjugate (FITC-BSA), Poly (styrene-4 sodium sulfonate) (PSS, MW = 70,000), poly (acrylic acid) (PAA, MW = 2000), poly (L-lysine hydrochloride) (PLL, MW=22,100), Fluorescein isothiocyanate PLL (FITC-PLL, avg. MW = 30,000) and poly (ethylene glycol) (PEG, MW=10,000) were purchased from Sigma. 4',6-diamidino-2-phenylindole (DAPI) in the form of VECTASHIELD, an aqueous mixture of DAPI in a glycerol-containing suspension, was purchased from Vector Labs. Trypan Blue was purchased from Gibco. Softisan 100 was a gift from Sasol GmbH. Water soluble, 50nm silica magnetite nanoparticles were purchased from Micromod GmbH as a 10 mg/mL suspension in PBS. Cadmium Selenide-Zinc Sulfide (CdSe) semiconducting nanocrystals (EviDots) in toluene were purchased from Evident Technologies. 10K MWCO regenerated cellulose (RC) and 100K MWCO cellulose ester (CE) 6 mL capacity floating dialysis columns (Float-a-Lyzer) were purchased from Spectrum Laboratories and dialyzed overnight against 4L double distilled water (Millipore) to remove sodium azide preservative prior to use.

Polymeric Nanoparticle Synthesis

To prepare polymeric NP, an aqueous phase was added to an equal volume of PLGA or PLA polymer dissolved in a binary solvent system. THF:Acetone,

MeCl:Acetone and MeCl:THF were selected as the binary solvent systems for polymeric NP synthesis. The volumetric ratio of the solvent pair was varied to investigate the effect of solvent polarity on nanoparticle formation. When surface functionalization was desired, the aqueous phase was supplemented with either a polyelectrolyte or a water-soluble polymer such as PEG, at a 0.5% concentration. Alternatively, the fluorescent dye DAPI was added to the aqueous phase at a 100 ug/mL concentration prior to phase inversion, after which the polymeric NP were dialyzed on 10K MWCO Float-a-Lyzers against double distilled water to remove non-entrapped dye.

Solid Lipid Nanoparticle Synthesis

To prepare SLN, Softisan 100 was dissolved in an anhydrous 80:20 mixture of NMP:Acetone as the binary solvent system at a concentration of 1%. To synthesize SLN bearing quantum dots (QD), 1 uL of a 1 mg/mL suspension of CdSe QD in toluene was added to the NMP/Acetone/Softisan mixture in a nitrogen-purged glove bag. An aqueous phase containing 1% PSS in ultrapure (100nm filtered) double-distilled water was prepared. To synthesize SLN bearing FITC-BSA/SPIO, the PSS-based aqueous phase was supplemented with 0.1% FITC-BSA and 100ug of silica-capped SPIO. The organic phase was rapidly infused into a scintillation vial containing the aqueous phase using a syringe accompanied by gentle shaking. Organic solvents were removed by dialysis against 4L ddH$_2$O in 10K MWCO RC Float-a-Lyzers in a low-speed stirring beaker for 2 hours. Next, excess polymer and FITC-BSA was removed by overnight dialysis against 4L ddH$_2$O in 100K MWCO CE Float-a-Lyzers with a complete change of dialysis buffer every 4 hours. For further surface functionalization of PLL, FITC-PLL was incubated at 4°C overnight with SLN-QD-PSS at a concentration of 1%, and dialysis against 4L ddH$_2$O using 100,000 MWCO CE Float-a-Lyzer was performed the following day to remove excess FITC-PLL.

Measurement of Polymeric and Solid Lipid Nanoparticle Hydrodynamic Radius and Zeta Potential

Polymeric NP were diluted by a factor of 15 in ddH$_2$O and adjusted to the desired pH for zetasizing analysis using either HCl and NaOH. Measurements of ζ and hydrodynamic radius were conducted in automatic mode on a Malvern Instruments Zetasizer 3000HS.

SLN were diluted by a factor of 50 in ddH$_2$O and adjusted to an acidic pH using HCl. Measurement of hydrodynamic radius and ζ was conducted using a Beckman-Coulter 440SX Zetasizer.

Fluorescence and Electron Microscopy of Polymeric and Solid Lipid Nanoparticles

Fluorescence microscopy of DAPI (VECTASHIELD)-loaded PLGA NP mounted in PBS, pH=7.4, was conducted on a Zeiss Axiophot equipped with bandpass excitation and emission filters specific for the dye. Digital photographs were acquired at optically- and digitally-enhanced 400X magnification and analyzed using Zeiss-supplied software. SLN-QD, SLN-QD with FITC-PLL surface functionalization, and SLN bearing FITC-BSA-SPIO were mounted in borate buffer, pH=8, and analyzed using a Nikon TE-2000U fluorescence microscope equipped with bandpass excitation and emission settings of 380/45ex, 585/40em for SLN-QD, and a standard FITC configuration to visualize dye-conjugated PLL on the SLN surface or entrapped FITC-BSA, at digitally and optically-enhanced magnification of up to 500X. For microscopy of FITC-BSA/SPIO bearing SLN, the solution was supplemented with 0.5% Trypan Blue to quench non-encapsulated FITC-BSA fluorescence. A 1.5T static magnet was placed on the right edge of the microscope slide to induce coordinated movement of FITC-BSA/SPIO SLN in response to the applied field. Images were analyzed using Image Pro Plus 5.1 software (Media Cybernetics).

SLN-QD were prepared for electron microscopy by mounting on Formvar grids with 50:50 EtOH:phosphotungstic acid solution to visualize solid lipid in negative relief using a Philips CM-12 electron microscope. Electron density of encapsulated QD enabled facile visualization.

X-ray Photoelectron Spectroscopy (XPS) of Polymeric NP

Polymeric NP were prepared for XPS surface analysis by dialysis of a 5 mL NP suspension against 500 mL 50% EtOH, flash freezing of NP in liquid nitrogen, followed by lyophilization for 48 h. The NP powder surface composition was analyzed using a Kratos Axis-Ultra X-ray photoelectron spectrometer equipped with a monochromatic Al Kα (1486eV) x-ray source, configured at 315W (25mA). Data was collected using a pass-energy of 40eV with 0.05eV steps. Elemental composition was determined using CasaXPS software.

Fluorescence Spectrophotometry of Solid Lipid NP

SLN-QD-PSS and SLN-QD-PSS-FITC/PLL were diluted 10-fold in borate buffer, pH=8. 2 uL of each suspension was then analyzed on a Nanodrop ND-3300 spectrofluorimeter configured for UV LED excitation and automatic sensitivity adjustment, with fluorescence emission intensity analyzed in the visible spectrum.

Acknowledgments

This work was supported in part by the Vanderbilt University Discovery Grant Program (VPS), and a National Eye Institute-Vanderbilt Vision Research Center Training Grant (AJ). We thank Michael V. Clark at Rohm and Haas for assistance with acquiring XPS data.

References

1. Sussman, E. M.; Jr., M. C.; Shastri, V. P. In *Functionalized Polymeric Nanoparticles*; Glembocki, O. J.; Hunt; C. E., Eds.; Materials Research Society, **2004**, p M12.9.
2. Muller, R. H.; Keck, C. M. *J Biotechnol* **2004**, *113*(1-3), 151-170.
3. Uner, M.; Wissing, S. A.; Yener, G.; Muller, R. H. *Pharmazie* **2004**, *59*(4), 331-332.
4. Wissing, S. A.; Kayser, O.; Muller, R. H. *Adv. Drug Deliv. Rev.* **2004**, *56*(9), 1257-1272.
5. Estey, T.; Kang, J.; Schwendeman, S. P.; Carpenter, J. F. *J. Pharm. Sci.* **2006**, *95*(7), 1626-1639.
6. Mohammed, A. R.; Weston, N.; Coombes, A. G.; Fitzgerald, M.; Perrie, Y. *Int. J. Pharm.* **2004**, *285*(1-2), 23-34.
7. Stroh, M.; Zimmer, J. P.; Duda, D. G.; Levchenko, T. S.; Cohen, K. S.; Brown, E. B.; Scadden, D. T.; Torchilin, V. P.; Bawendi, M. G.; Fukumura, D.; Jain, R. K. *Nat. Med.* **2005**, *11*(6), 678-682.
8. Woodle, M. C.; Lasic, D. D. *Biochim. Biophys. Acta* **1992**, *1113*(2), 171-199.
9. Guggi, D.; Marschutz, M. K.; Bernkop-Schnurch, A. *Int. J. Pharm.* **2004**, *274*(1-2), 97-105.
10. Diamandis, E. P.; Christopoulos, T. K. The biotin-(strept)avidin system: principles and applications in biotechnology. *Clin. Chem.* **1991**, *37*(5), 625-636.
11. Finnemann, S. C.; Bonilha, V. L.; Marmorstein, A. D.; Rodriguez-Boulan, E. *Proc. Natl. Acad. Sci. USA* **1997**, *94*(24), 12932-12937.

Chapter 9

Construction of Multifunctional DDS:Transferrin-Mediated Tat and Drug-Loaded Magnetic Nanoparticles

Jie Huang[1], Peng Yao[1], Aijie Zhao[1], Chunshang Kang[1], Peiyu Pu[2], and Jin Chang[1]

[1]Institute of Nanobiotechnology, School of Materials Science and Engineering, Tianjin University, Tianjin 300072, People's Republic of China
[2]General Hospital, Tianjin Medical University, Tianjin 300072, People's Republic of China

This paper describes a new formulation of magnetic nanoparticles coated by a novel polymer matrix----O-Carboxylmethylated Chitosan (O-CMC) as drug/gene carrier. The O-CMC magnetic nanoparticles were derivatized with a peptide obtained from the HIV-tat protein and transferrin to improve the translocation function and cellar uptake of the nanoparticles. To evaluate the O-MNPs-TAT-Tf system as drug carriers, Methotrexate (MTX) was incorporated as a model drug and MTX-loaded O-MNPs-TAT-Tf with an average diameter of 75nm were prepared and characterized by TEM, AFM and VSM. The cytotoxicity of MTX-loaded O-MNPs-TAT-Tf was investigated with C6 cells. The results showed that the MTX-loaded O-MNPs-TAT-Tf could be a novel magnetic targeting carrier. The ability of O-MNPs-Tat-Tf crossing BBB in rats was also investigated by single photon emission computed tomography (SPECT).

It is well known that there has been great interest in developing and testing of iron oxide nanoparticles for tumor detection and therapy over the past two decades. In brain research, nanodispersed iron oxides have been used as carriers of diagnostic and therapeutic agents for mapping of blood-brain barrier (BBB) disruption to improve tumor detection and therapy (1-5).

We hypothesized that a superparamagnatic iron oxide attached with a membrane translocation signal (MTS) peptide might be capable of producing high level of cell internalization. Several MTS have been described including the third helix of the homeodomain of Antennapedia, the peptide derived from anti-DNA monoclonal antibody, VP22 herpes virus protein and HIV-1tat peptide. HIV-1 tat peptide is an 86 amino acid polypeptide and is essential for viral replication. It has been found to be capable of traveling through cellular and nucleic membranes freely. Its membrane translocation function is dominated by a short signal peptide, namely GRKKRRQRRR (amino acid residues48-57). More recently, it has been demonstrated that the core peptide itself rather than the entire protein is capable of translocating a variety of molecules, including fluorescent probes, peptides, pathogenic epitopes, and proteins (6).

Since transferrin was discovered more than half a century ago, a considerable effort has been made towards understanding transferrin-mediated iron uptake. Apart from iron, many other metal ions of therapeutic and diagnostic of interest can also bind to transferrin at the iron sites and their resultant transferrin complexes can be recognized by many cells. Therefore, transferrin has been thought of as a delivery system into cells. In addition, it has also been widely applied as a targeting ligand in the active targeting of anticancer agents, proteins and genes to primary proliferating cells via transferrin receptors (7, 8).

Here we designed and prepared a novel polymer (O-CMC) coated magnetic nanoparticles conjugating with Tat and Transferrin as drug/gene carrier. We chose O-CMC as the coating agent because O-CMC is biocompatible, biodegradable, nontoxic and water soluble (9, 10) and also, because it has some unique antitumor and antibacterial bioactivities (11). Another appealing characteristic of O-CMC is its bifunctional groups-carboxyl and amino groups, which could be readily covalently coupled with diverse bioactive macromolecules, anticancer drugs and liposomes.

Materials and Methods

Materials

O-CMC (Mw=40000) was made by our own in the Institute of Nanobiotech.,Tianjin University. Tat-peptide was synthesized by GL Biochem (Shanghai) Ltd. Transferrin, Methotrexate (MTX), N-succinimidyl-3-(2-

pyridyldithio) propionate (SPDP), diethylenetriaminepentaacetic acid (DTPA),1-ethyl-3-(3-dimethylaminopropyl)- carbodiimide (EDC), 3-(4,5-dimethylthiazol-2-yl)- 2,5-diphenyltetrazolium bromide (MTT) and gel filtration were obtained from Sigma, USA. Ferric chloride hexahydrate, ferrous chloride tetrahydrate, sodium hydroxide (NaOH), hydrochloric acid (HCl), stannous chloride ($SnCl_2$) and dimethylsulphoxide (DMSO) which were all analytic purifications were obtained from Kewei biotech (Tianjin. China). Double distilled water was used for all the experiments.

Synthesis of O-CMC Magnetic Nanoparticles (O-MNPs)

SolutionA: 3.0g O-CMC(MW=40,000)+0.326g Ferric in 15.0ml H_2O was passed through a filter (220nm) and pre-chilled on ice.
Solution B: 0.150g ferrous in 1ml H_2O was passed through a filter (220nm) and pre-chilled on ice.
Solution B was added dropwise into solution A with vigorous stirring, followed by addition of 5ml 20% NaOH. After that, the solution was slowly heated to 70-90°C within one hour. The solution was kept at this temperature and stirred for another 30mins. After elimination of sodium hydroxide by dialysis (MWCO 14,000) in distilled water, O-MNPs were separated by a magnet and the pellets were prepared using lyophilization
For O-MNPs, a variable amount of the O-CMC (ranging from 0.3 to 4g), a variable proportion of Fe^{3+}/Fe^{2+} (ranging from 4:1 to 1:4), and a variable concentration of NaOH (ranging from 5% to 50%) was respectively designed and we obtained the optimized conditions as described above.

Tat and Transferrin Conjugation via a Disulfide Linkage and Drug Loading

Synthesis of 2-Pyridyl Disulfide Derivatized O-MNPs

The O-MNPs were reacted with SPDP to yield a 2-pyridyldithiol-end group on the surface. In brief, 2 ml of O-MNPs (20mg/ml) in 0.1 M phosphate buffer (pH=7.4) was mixed with 2.5 ml of SPDP in DMSO (20mM, 0.05mmol). The mixture was incubated at room temperature for 60min. Surplus SPDP were removed by gel filtration. The final void volume of 4.4ml was recovered.

Synthesis of O-MNPs-TTat-Tf

One ml of Tat in 0.1 M phosphate buffer pH 7.4(5mg/ml) was added to 1 ml of 2-pyridyl disulfide derivatized O-MNPs. The mixture was allowed to stand overnight at room temperature to form a disulfide linkage between the surface of O-MNPs and Tat (*12*). The resultant solution was applied to a magnetic field to separate O-MNPs-Tat conjugates equilibrated in 0.1 M phosphate buffer(pH 7.4) and the excluded volume containing free Tat was saved. 1mg of Tf in 0.5ml PBS (pH=7.4) was asdded to 1ml O-MNP-Tat solution, then 2ml EDC was added in the system. The mixture was allowed to stand 4h at 4°C.The resultant solution was purified by a magnetic field to obtained O-MNPs-Tat-Tf. In order to measure the number of Tat and transferrin attached, the free Tat and transferrin was quantitated using a standard peak value of Tat (5mg/ml), prepared by using high performance liquid chromatogram (HPLC).

MTX Loading To O-MNPs-Tat-Tf

To evaluate the O-MNPs-Tat -Tf as drug carriers, MTX was incorporated as a model drug, and MTX-loaded O-MNPs-Tat-Tf were prepared. In brief, 5mg MTX was dissolved in 5ml 0.1 M phosphate buffer (pH=7.4) and mixed with 1ml O-MNPs-Tat-Tf solution in PBS (pH7.4). 1 ml of EDC solution was added to this mixture. The final mixture was allowed to stand for 2 h at room temperature in the dark and then dialyzed against distilled water using MWCO 12000g/mol dialysis membrane. The medium was replaced every hour for the first 8 h and then every 3 h for 1 day. The MTX-loaded O-MNPs-Tat-Tf were separated under magnetic field and then freeze-dried. The amount of MTX attached was determined quantitating the free MTX using UV spectrophotometer at 372nm.

^{99m}Tc label to O-MNPs- Tat-Tf

One ml O-MNPs-Tat-Tf O-MNPs-Tat-Tf in 1ml 0.1 M phosphate buffer (pH 7.4) react with 5mg DTPA in room temperature for 1h. And unconjugated DTPA was separated by applying resultant solution to a magnetic field. Then the O-MNPs- Tat -Tf-DTPA was added 5 mg $SnCl_2$ followed immediately by addition of 0.175 ml HCl (37%). 0.5 ml $^{99m}TcO_4^{-}$[25mCi] to the above mixture kept in room temperature for 20min.

In vivo Studies

Rats under anesthesia were injected in the tail vein with radio labeled O-MNPs-TAT-Tf. Then the rats were kept between the poles of magnet (1T) for 30 min and studied by single photon emission computed tomography (SPECT).

Building of Simulated Model and In-vitro Investigation of MTX-loaded O-MNPs-Tat-Tf

The magnetic responsiveness of MTX-loaded O-MNPs-Tat-Tf was determined by a simulated model, and the influence of flow rate and magnetic field were investigated. Figure 1 shows the Blood Circulation Simulated Model where only one branch of silicone tube is used to simulate the simplest blood vessel network. About 40 ml of the MR fluid is placed in a beaker as the source fluid for the experiment. The fluid is drawn by a peristaltic pump and pushed through the tube. The tube then passes a magnetic pole and reaches the collector beaker sitting on top of a scale to weigh any leaking MTX-loaded O-MNPs-Tat-Tf through the magnet. The inner diameter (D) of the tube was 0.8 mm from the source beaker to the collector beaker. The tube size at the pole was much larger than arterioles (0.07mm) found in the human body. The flow rate Q was controlled by the peristaltic pump which was set between 0.8 to 4.3 ml/min corresponding to a flow velocity from 4 to 21.5 mm/s inside a 0.8 mm tube. Even the smallest Q was still higher than the typical flow rates found in human arterioles. Therefore, if the MTX-loaded O-MNPs-Tat-Tf can be blocked in our model, it should work better for vessels of the size of human arterioles. A portable (dimension: 25×10×16cm and weight: 5 kg) permanent magnet assembly was used to generate a magnetic field up to 2.5T. A pair of neodymium columned magnets was mounted on a stage with two pieces on each side (dimension 3.14cm^2×5cm per piece and field strength: 2T/piece). The two sets of magnets were connected by two cone-shaped iron poles which concentrated the magnetic field between their tips (10mm in diameter, 60 ° angles). The field strength was controlled by manually adjusting the gap width between the poles. The magnetic responsiveness of

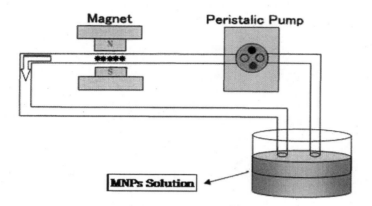

Figure 1. A Simulated Model for Magnetic Responsiveness Testing

158

nanoparticles was determined as follows. Approximately 0.1g of magnetic nanoparticles were suspended in a carrier fluid made of 20% O-CMC and 5% sodium chloride (which is a kind of blood supplement) to simulate the human blood. The viscosity of the carrier fluid was the same as whole human blood. The silicone tube was in direct contact with one pole. The gap between the two poles was changed respectively from 5mm to 30mm; essentially only one pole is used. The magnetic field strength (H) was altered from 2T to 0.35T. This one pole geometry is closer to most clinical situations where tumors may not always be accessible to two poles. An electro-scale (BS210S, Beijing) was used to measure the weight (W) of the leaking O-MNPs-Tat-Tf as a function of time. This data was taken every change and sent to the computer.

Antitumor Test In Vitro

Each U-937 human lymphoma and C6 brain glioma cell line (5×10^4cells/ml) was plated in 24- well plates and was incubated at 37°C. After 24h, cells were treated with MTX, MTX-O-MNPs-Tat, MTX-O-MNPs-Tf, MTX-O-MNPs-Tat-Tf (final concentration 0.1mg/ml). Blank MNPs were used as controls. Microplates were incubated at 37°C in a humidified atmosphere of 5% CO_2 48h. Then 0.03ml of 5mg/ml MTT solution was added to each well and the plates were incubated at 37°C for 4h, then to each well plate was added 0.5% DMSO. The cytotoxicity was measured with colorimetric assay based on the use of MTT (*13*). The results were read on a multiwell scanning spectrophotometer (Multiscan reader), using a wavelength of 570 nm. Each value was the average of 6 wells (standard deviations were less than 10%).

Evaluation of Ability of O-MNPs-Tat-Tf Crossing BBB inVivo

^{99m}Tc label to O-MNPs-Tat-Tf

One ml O-MNPs-Tat-Tf in 1ml 0.1 M phosphate buffer (pH 7.4) react with 5 mg DTPA in room temperature for 1h. The resultant solution was applied to a magnetic field to separate unconjugated DTPA. Then the O-MNPs-Tat-Tf-DTPA was added 5 mg $SnCl_2$ followed immediately by addition of 0.175 ml HCl (37%). 0.5 ml $^{99m}TcO_4^-$[25mCi] to the above mixture kept in room temperature for 20min.

In Vivo Studies

Rats (Male Fisher, 225-250g, which were supplied by Tianjin Medical University) were anesthetized with an intraperitioneal injection of a 3:2:1 (v : v : v) mixture of ketamine hydrochloride(100mg/ml), acepromazine maleate(10mg/ml) and xylazine hydrochloride(20mg/ml) at a dose of 0.1mg/100g body weight. Then the rats were injected in the tail vein with radio labeled O-MNPs-Tat-Tf about 1ml. After the injection the rats were kept between the poles of magnet (1T) for 30 min and studied by single photon emission computed tomography (SPECT).

Results and Discussion

Synthesis and Characterization of O-CMC Magnetic Nanopartilces

Effect of O-CMC Concentration

From Table I it could be deducted that the nanoparticle size minished with increased O-CMC concentration. This may be due to the fact that O-CMC coats prevent magnetic particles reuniting so as to obtain particles with nano size. However, the nanoparticles size increased when the O-CMC concentration maintains increasing. Thismay be due to the combining capacity of O-CMC to iron. When O-CMC concentration increased continuously, the combining capacity of O-CMC to iron enhanced either. The preparation procedure gives priority to the combining capacity when the O-CMC concentration exceeded 50%, resulting in larger size nanoparticles.

Table I. Effect of O-CMC Concentration.

Sample	O-CMC concentration(w/v)	Average Diameter
S1	3%	100nm
S2	10%	70nm
S3	15%	45nm
S4	25%	35nm
S5	40%	20nm
S6	60%	40nm

Effect of NaOH Concentration

As seen in Table II, in the first, the particles size decreased with increased concentration of NaOH. It was because Na^+ adsorbed to the surface of the nanoparticles played an important role in separating formed nanoparticles. And then, with NaOH concentration increased to a certain extent, on the contrary, the size of nanoparticles increased. The result can be explained by precipitation capacity of NaOH. Increasing the NaOH concentration increased the precipitation capacity of NaOH, which is propitious to prepare larger nanoparticles.

Table II. Effect of NaOH Concentration.

Sample	NaOH concentration(w/v)	Average Diameter
S1	5%	90nm
S2	15%	70nm
S3	20%	45nm
S4	30%	35nm
S5	50%	50nm

Effect of Fe^{3+}/Fe^{2+} Molar Rate

The effect of the Fe^{3+}/Fe^{2+} molar rate on nanoparticles size is shown in Table III. It can be seen that the size of nanoparticles increases when the molar rate is larger than 2:1 or smaller than 2:3. Smaller nanoparticles can be prepared while Fe^{3+}/Fe^{2+} mol rate is between the two above limits. The reason is very clear that the Fe^{3+}/Fe^{2+} mol rate in Fe_3O_4 is 2:1, and smaller stable nanoparticles could be obtained as long as the mol rate is close to 2:1 while excessive Fe^{2+} is needed (a part of Fe^{2+} would be lost due to oxidation).

Table III. Effect of Fe^{3+}/Fe^{2+} Molar Rate.

Sample	Fe^{3+}/Fe^{2+} mol rate	Average Diameter
S1	4:1	80nm
S2	2:1	120nm
S3	1:1	20nm
S4	2:3	50nm
S5	1:4	75nm

Effect of Reaction Temperature

Table IV shows the effect of reaction temperature. The nanoparticle size decreased with increased reaction temperature. Increasing the reaction temperature enhanced both the rate of adsorption of O-CMC and the viscosity of the coat phase. All these factors would reduce the extent of aggregation of the nuclei and reduce the particle size. However the prepared O-MNPs are not stable when the temperature exceeds 90°C.

Table IV. Effect of Reaction Temperature

Sample	Reaction temperature(°C)	Average Diameter
S1	40	130nm
S2	60	70nm
S3	80	55nm
S4	100	50nm

Effect of Stirring Rate

As seen in Table V, the nanoparticle size decreased with increased stirring rate. This may be due to the fact that high stirring speed could generate powerful shearing strength so as to prevent nanoparticles reuniting. At the same time, too high stirring speed such as 2400rmp or above is not necessary in order to avoid splashing of reaction solution.

Table V. Effect of Stirring Rate

Sample	Stirring rate (rmp)	Average Diameter
S1	400	150nm
S2	700	110nm
S3	1300	100nm
S4	1700	85nm

Characterization of O-CMC Magnetic Nanoparticles

The morphology, average size and size distribution of the MTX Loaded O-MNPs-Tat-Tf were measured by JEOL-100CXII Transmission Electronic Microscopy (TEM), MMAFM/STM+D3100 Atomic Force Microscopy (AFM), and 90 Plus/BI-MAS Multi Angle Particle Sizing Instruments (Brookhaven Instruments Co.) The results are shown in Figure 2, Figure 3. and Figure 4. The average size of MTX Loaded O-MNPs-Tat-Tf is about 75nm with well-shaped morphology and narrow size distribution.

HPLC Analysis of O-MNPs-Tat-Tf

Figures 5 and 6 show the standard sample of Tat (0.2mg/ml) and transferrin (0.24mg/ml).We can find the peak of Tat and transferrin in the two Figures. Figure 7 shows the supernatant fluid of O-MNPs-Tat-Tf. There were no identical peaks in Figure as in Figure 5 and 6. So from the HPLC analysis we found that the Tat and transferrin were conjugated with O-MNP successfully.

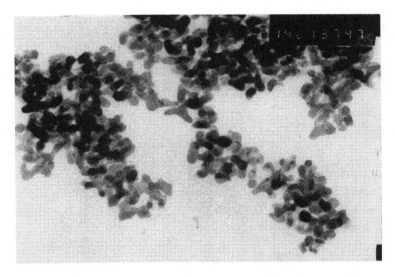

Figure 2. TEM Photograph of MTX Loaded O-MNPs-Tat-Tf (×103.5K)

Figure 3. AFM Photograph of MTX Loaded O-MNPs-Tat-Tf

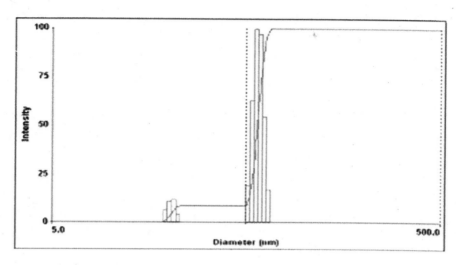

Figure 4. Size Distribution of MTX Loaded O-MNPs-Tat-Tf

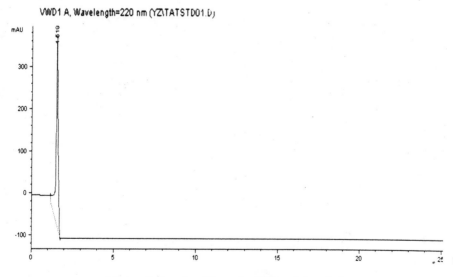

Figure 5. Standard Sample of Tat (0.2mg/ml)

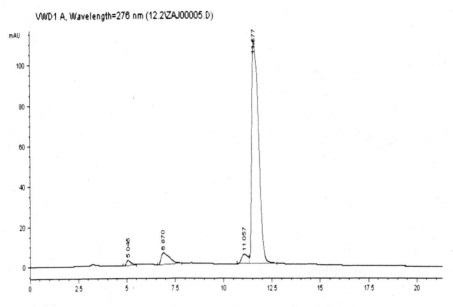

VWD1 A, Wavelength=276 nm (12.2\ZAJ00005.D)

Figure 6. Standard Sample of Transferrin (0.24mg/ml)

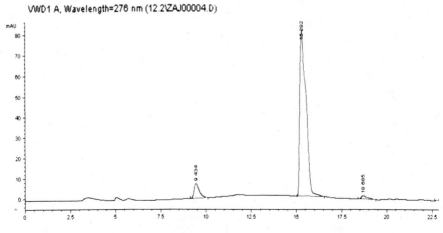

VWD1 A, Wavelength=276 nm (12.2\ZAJ00004.D)

Figure 7. Supernatant Fluid of O-MNPs-Tat-Tf

Magnetic Properties of MTX-loaded O-MNPs-Tat-Tf

Figures 8 and 9 show magnetic responsiveness of MTX Loaded O-MNPs-Tat-Tf, tested by a simulated model. The superparamagnetic property of MTX Loaded O-MNPs-Tat-Tf was determined by VSM (Vibrating Samples Magetometer, LDJ9600-1). Figure 8 shows the effect of magnetic field on the assembling capability of magnetic nanopartlces. Model (1) is the regression formula of this curve. Figure 9 shows the influence of flow rate on assembling capability of magnetic nanopartiles. Model (2) is the regression formula of this curve. From these results, we could calculate the assembling condition of the magnetic nanoparticles. The assembling condition of magnetic field should come up to 1.2T or more. The condition of flow rate should be less than 0.626cm/s. All of the two conditions confirm that the magnetic nanoparticles can freely carry and stably assemble in capillary vessels of bodies.

$$CR = -16.68469+0.01064H-2.74006E-7H2 \ (R^2=0.9809) \qquad (1)$$

In the formula, CR represents conglomeration rate, H represents magnetic intensity.

$$CR = 113.18785-21.50469V + 1.23725 \ V^2 \ (R^2=0.9844) \qquad (2)$$

In the formula, CR represents conglomeration rate, V represents flow rate.

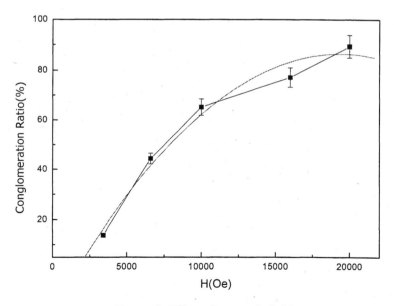

Figure 8. Effect of magnetic field

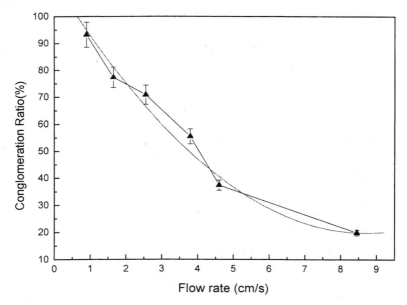

Figure 9. Effect of flow rate

The magnetization curve for MTX Loaded O-MNPs-Tat-Tf is shown in Figure 10, and was determined through VSM. The remnant magnetization (Mr) value for the products was 0.005725 emu/g, which was 0.08% of the saturation magnetization (Ms=7.247emu/g). The result is perfectly comparable to that reported in the literature (*14*). The value of the coercive force (Hc=1.195Oe) of MTX-loaded O-MNPs-Tat-Tf, which is close to zero, is evidence that the MTX-loaded O-MNPs-Tat-Tf we prepared have the magnetic characteristics that are close to the superparamagnetic matter.

Unique Characteristics of O-MNPs-Tat-Tf

We originally prepared O-CMC coated magnetic nanoparticles conjugated with Tat and transferrin as drug carrier. The carrier has some unique characteristics as follows.

1. The coating agent O-CMC is biocompatible, biodegradable, nontoxic and water soluble.

2. O-CMC also has some unique antitumor and antibacterial bioactivities.

3. Another appealing characteristics of O-CMC is its bifunctional groups (-NH$_3$,-COOH) which provides great scope for binding to diverse bioactive macromolecules, anticancer drugs etc.

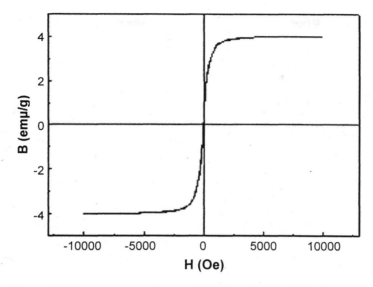

Figure 10. Magnetization Curve of MTX Loaded O-MNPs-Tat at 25°C. Data from the curve were summarized as below:

Parameters	Value
Hmax	9961Oe
Hc	1.195Oe
Hk	626.6Oe
Br	-0.005725emu/g
Bs	7.247emu/g

4. O-CMC coated round the magnetic nanoparticles is a cationic polymer, which could enable the magnetic carrier system to bind gene for gene therapy.

5. In this paper, Tat was manipulated as a molecular tool for its unique translocation property. Tat can leave cells from which it is synthesized and cross the membrane of adjacent cells, where it localizes in the nucleus. Importantly, Tat maintains its activity and, once inside the nucleus, is able to trans-activate a number of genes and thus can modulate certain cellular activities. The ability of Tat to translocate across the cell membrane is of major importance, particularly since the cell membrane can be a formidable obstacle for the magnetic nanoparticles to negotiate; therefore, by manipulating this property of tat, non-permeable magnetic nanoparticles can be introduced into the cell.

6. Transferrin has been widely applied as a targeting ligand in the active targeting of anticancer agents, proteins and genes to primary proliferating cells via transferrin receptors. Therefore, transferrin has been thought of as a 'delivery system' into cells.

In Vitro Antitumor Activity Test

The results of the in vitro antitumor activity of free MTX, MTX-O-MNPs and blank O-MNPs after incubation with U-937 human lymphoma and C6 brain glioma cells are shown in Figure 11 and Figure 12. (1) MTX-O-MNPs showed the similar antitumor effect compared to the free MTX in different cancer cells. (2) As time passed by, the cell inhibition effect of MTX-O-MNPs gradually increased,which proved that it possessed good antitumor effect. Meanwhile, inhibition effect of free MTX reached its maximum value after 72 h.

Figure 13 shows the antitumor activity of MTX, MTX-O-MNPs-Tat, MTX-O-MNPs-Tf, MTX-O-MNPs-Tat-Tf and blank O-MNPs under the magnetic field comparing with Figure 14. (1) MTX-O-MNPs-Tat-Tf showed the similar antitumor effect as Figure 12 (2) We find that Figure 13 shows more antitumor ability than Figure 14. That may be the magnetic nanparticles under the magnetic field could more easily combine with cells. Compared with the inhibition effect curve of free MTX, the inhibition effect curve of MTX-O-MNPs is more assuasive indicating that MTX-O-MNPs have obvious controlled drug release function. (3) Meanwhile, it also can be seen in the Figure that O-MNPs have little cytotoxicity on the tumor cells and barely exerted any effect on the growth of the cells. This demonstrated that O-MNPs have good biocompatibility. (4) The MTX-O-MNPs-Tat-Tf showed a little higher antitumor effect than MTX-O-MNPs, while still retaining controlled drug release characteristics; this may be due to the membrane translocation property of tat and the targeting ability of transferrin, producing high levels of cell internalization.(5) The curve of Figure 13 more centralize than Figure 13. We think it may be the effect of magnetic field. The nanoparticles under magnetic field could combine with cancer cells more efficiently.

Ability of Crossing BBB for O-MNPs-Tat-Tf in Vivo

Figure 15 is the SPECT of O-MNPs-Tat-Tf in rats. The main focus of this study was to examine whether the O-MNPs-Tat-Tf could pass the blood-brain barrier (BBB) in vivo. We know that the technetium-99m alone can not pass the BBB. After it was connected with O-MNPs-Tat-Tf we found that there were radiotracers in the rats brains. So it illustrated that the O-MNPs-Tat-Tf could pass the BBB under the magnetic field. Therefore the O-MNPs-Tat-Tf could be used as drug or gene delivery for the treatment of brain diseases. Tat has special ability to cross the biomembrane. It may help the particles to cross the BBB. From the picture not only in brains but also in thoracic cavity has the radiotracer, that maybe the effectors of transferrin. There are many cells and organs in the rats have transferrin receptors, the particles could easily combine with such cells.

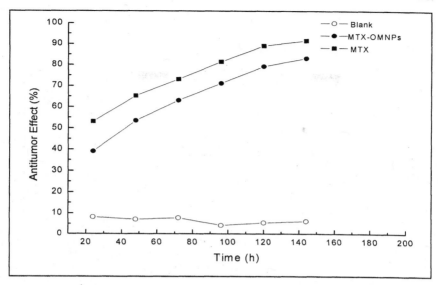

*Figure 11. Antitumor effect of free MTX,, MTX-O-MNPs and blank
O-MNPs on U-937 human lymphocytes*

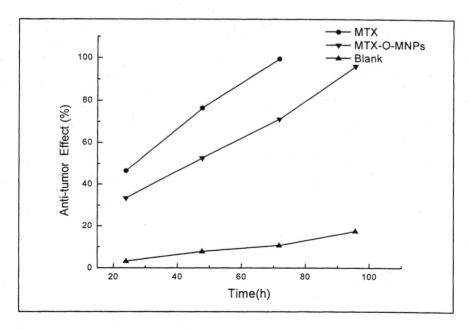

*Figure 12. Antitumor effect of free MTX,, MTX-O-MNPs and blank
O-MNPs on C6 brain glioma cells*

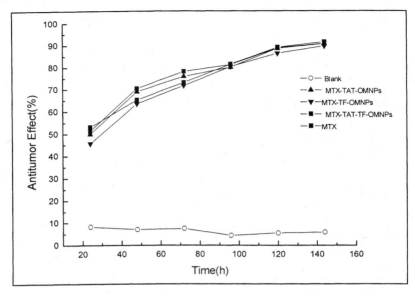

Figure 13. Antitumor effect of free MTX, MTX-O-MNPs-TAT, MTX-O-MNPs-Tf, MTX-O-MNPs-Tat-Tf and blank O-MNPs on C6 brain glioma cells (with magnetic field)

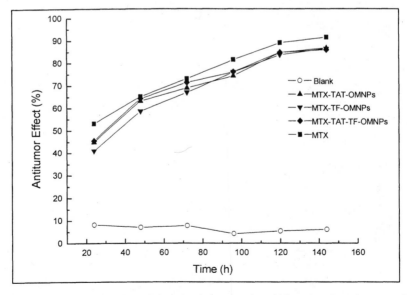

Figure 14. Antitumor effect of free MTX, MTX-O-MNPs-Tat, MTX-O-MNPs-Tf, MTX-O-MNPs-Tat-Tf and blank O-MNPs on C6 brain glioma cells (without magnetic field)

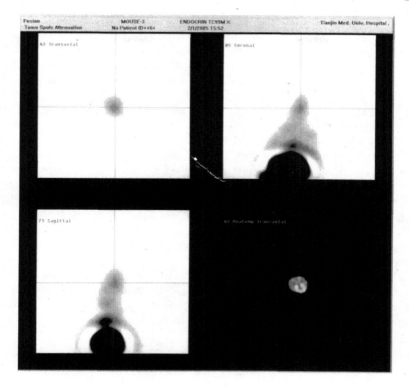

Figure 15. SPECT of O-MNPs-Tat-Tf in rats. Down-left and up-right was the distribution of O-MNPs-Tat-Tf in rats, in which the bright point of cross line was the area of the radioactivity aggregation of O-MNPs-Tat-Tf in the brain of the rat. Up-left and down-right was the amplification of the area of the radioactivity aggregation of O-MNPs-Tat-Tf in the brain of the rat under different background.

In addition there are many transferrin receptors on the surface of the tumor cells, so the O-MNPs-Tat-Tf may still stain at tumor tissue and enter into the tumor cells through the receptor-mediated mechanism in absence of magnetic field.

Conclusions

We originally used the biocompatible, biodegradable and nontoxic O-CMC as a novel polymer matrix to prepare superparamagnetic nanoparticles of an average diameter of 75nm, aimed at testing whether O-CMC could be a versatile polymer matrix of the suerparamagnetic nanoparticles as drug or gene carrier.

172

The conjugation of tat peptide and transferrin to the surface of the O-MNPs showed that a stable and reproducible formulation has been obtained suggesting that O-MNPs-TAT-Tf could be a novel targeting carrier.

The in vitro simulated model investigation confirmed that the magnetic nanoparticles can freely carry and stably assemble in capillary vessels of bodies. In the in vitro antitumor activity test, The MTX-O-MNPs-TAT-Tf showed better antitumor effect than MTX-O-MNPs while still possessed good controlled drug release characteristic.

The desired nanoparticles for loading DNA were obtained in order to realize and optimize the efficiency of gene therapy for tumors. The result shows that the O-MNPs could successful transfect gene to the cell. Studies *in vivo* demonstrated that the O-MNPs-Tat-Tf were successful located in brain of the rats by photography of SPECT. The O-MNPs-Tat-Tf successfully passed the BBB under the magnetic field. Therefore the O-MNPs-Tat-Tf is promising for therapy of tumors in the brain, such as glioma. It also has potential application in other brain diseases. Our studies in human glioma cells also showed effectively and used as a gene therapy optional carrier

We hypothesized that the combination of the magnetic targeting characteristic with the translocation property and high cellular uptake of Tat could be a better therapy. We believe that this exciting technology can offer a creative advance in therapeutic drug and gene delivery .

Acknowledgment

This work was performed under the auspices of National Nature Science Foundaton (50373033) and Key Project Foundation from Tianjin Science and Technology Committee (05YFJZJC01001),to which the authors wish to express their thanks.

References

1. Chang, J. *Chinese Journal of Biomedical Engineering* **1996**, *15*(4), 354-359.
2. Chang, J. *Chinese Journal of Biomedical Engineering* **1996**, *15*(2), 97-101.
3. Fricker, J. *DTT*, **2001**, *6*(8).
4. Häfeli, U., Pauer, G., Failing, S., Tapolsky; G. *Journal of Magnetism and Magnetic Materials*, **2001**, *225*, 73-78.
5. Pulfer, S.K., Gallo, J.M. In: *Scientific and Clinical Applications of Magnetic Carriers*; Hafeli, U. et al. Eds.; Plenum Press: New York, 1997; p. 445.
6. Xe, W., Xu, P., Wang, W., Liu, Q. *Carbohydrate Polymers* **2002**, *50*, 35-40.

7. Moore, A., Josephson, L., Bhorade, R.M., Basilion, J.P.,Weissleder, R. *Radiology* **2001**, *221*, 244-250.

8. Bulte, J. M. W., Zhange, S. C., van Gelderen, P., Herynek, E. K., Duncan, I. D., Frank, J. A. *Proc. Natl. Acad. Sci. U.S.A.* **1999**, *96*, 15256-15261.

9. Begona, Carreno-Gomez Ruth Duncan. *International Journal of Pharmaceutics* **1997**, *148*, 231-240.

10. Liu, X. F.; Guan, Y. L.; Yang, D. Z.; Yao, K. D. *Journal of Applied Polymer Science*, **2001**, *79*(7), 1324-1335.

11. Hermanson, G. T. In: *Bioconjugate Techniques Academic Press:* San Diego, CA 1996; pp. 56-60.

12. Deshpande, A.; Toledo-Velasquez, D.; Wang, L. Y. et al., *Pharm. Res.* **1994**, *11*(8), 1121-1126.

13. Mosmann, T. *J. Immunol. Methods* **1983**, *65*(1-2), 55-63.

14. Lewin, M., Carlesso, N., Tung, C.H., et al., *Nature Biotechnology* **2000**, *18*(4), 410-414.

Chapter 10

Encapsulation of Essential Oils in Zein Nanospherical Particles

Nicholas Parris, Peter H. Cooke, Robert A. Moreau, and Kevin B. Hicks

Eastern Regional Research Center, Agricultural Research Service, U.S. Department of Agriculture, 600 East Mermaid Lane, Wyndmoor, PA 19038

Essential oils, oregano, red thyme, and cassia (100% pure oil), were encapsulated by phase separation into zein particles. Typical yields were between 65% and 75% of product. Encapsulation efficiency of all oils was 87% except for cassia oil which was 49%. Loading efficiency of all oils was 22% except for cassia oil which was 12%. Topographical images indicated that the powders were made up of irregularly shaped particles (~50 μm) containing close-packed nanospheres. Approximately 31% oregano encapsulated nanospheres, had mean diameters greater than 100 nm compared to 19% for the zein alone nanospheres. Examination of the zein-oregano oil and zein alone powder particle surfaces by scanning electron microscopy and atomic force microscopy revealed areas of close-packed spheres. Efforts to localize oil in powder particles by fluorescence from Nile Red stain using confocal laser scanning microscopy were equivocal because the zein particles fluoresced at equal intensities as the zein-oregano oil powdered particles. Enzymatic in vitro digestion of zein particles with pepsin at a concentration ratio of 10:1 was complete after 52 h in phosphate-citrate buffer, pH 3.5, at 37 °C. However, less than half the particles were digested after

4 h, the typical gastric emptying time for humans and some monogastric animals. Nonenzymatic, aqueous in vitro release of essential oils from encapsulated zein particles was carried out in phosphate buffered saline at pH 7.4 and 37 °C. Very little oregano appeared to coat the particle surface because no initial burst of oregano was released which indicated that the oregano oil is encapsulated or entrapped within the particle. Release occurred at varying rates over 20 h probably from different locations within the closely packed nanospheres of different sizes. Gel electrophoresis SDS-PAGE of zein incubated with freeze-dried swine manure solids at 37 °C indicated that preformed microbial enzymes capable of digesting zein within minutes were present in the manure. Except for differences in size of nanospheres, no structural differences were resolved by several microscopic distribution methods, suggesting that the oil and protein phases were blended during phase separation.

In light of the rapid evolution of bacterial resistance to multiple drugs resulting from overuse of antibiotics in humans and livestock, the development of new antimicrobial compounds is imperative. The antimicrobial activity of plant oils and extracts has been recognized for many years. Grinding of spices has been shown to break down the secretory cells, glands and other tissues in which the, oils are located (1). Essential oils are slightly soluble in water and impart their odor and taste to the water. They contain terpenes, alcohols, esters, aldehydes, ketones, phenols, ethers, and other minor compounds. Essential oils have a wide spectrum of biological activities, including growth inhibition observed against bacteria, yeasts, and fungi (2). Bose et al. (3, 4) measured the bactericidal efficiency of several essential oils and found their effects to be high against Gram-negative bacteria but very low against Gram-positive organisms. In addition, they found that essential oils containing aldehydes were more effective than those containing alcohols. The mechanism of action is not completely understood, however, essential oils containing terpenoids and phenolics are thought to act against microorganisms through membrane disruption (5). Thyme, cinnamon, mustard, oregano, and rosemary have been shown in numerous studies to be highly antimicrobial (6).

To have an efficient delivery system it must be capable of administering the essential oil to a specific site under a myriad of conditions with minimal loss of oil. In addition it is generally recognized and desirable that delivery systems be prepared from natural materials if possible since they tend to be more biocompatible and less likely to exhibit undesirable interactions with the host compared to synthetic compounds. In this regard proteins are desirable since they possess unique functional properties and easily form gels and emulsions, which are suitable for the encapsulation of bioactive compounds like the essential oils. Delivery systems have been developed by the pharmaceutical and biomedical industries that trap molecules of interest within well defined networks (7, 8). In addition, because of the availability of physiochemical information for most food proteins the development of novel delivery systems in the form of hydrogels, micro- or nano-particles from these compounds is possible. In the 1980's, delivery systems were primarily limited to protein microparticles which found wide application in the food industry. More recently, nanoparticles (less than 100 nm in length) offered a promising means of delivering poorly soluble compounds due to their subcellular size. Previously, we have shown that essential oils can rapidly be encapsulated into zein nanospheres by phase separation (9). Corn zein is a prolamine rich protein that functions as the major storage protein in corn. It is found in protein bodies in the endosperm of the corn kernel and because of its hydrophobic properties is widely used for films and coatings and is suitable for encapsulation of a variety of compounds. Corn zein microspheres have been researched as drug and vaccine carriers that could increase the immune response of co-administered immunogens and as a delivery system for ovalbumin as a parenteral antigen (10), ivermectin, a highly effective parasiticide for use in farm animals (11), antitumor drugs for delivery into the tumor-feeding arteries (12), abamectin, a light-sensitive lactone used to control pests (13), drugs using a solvent-evaporation process for the preparation of injectable controlled-release microcapsules (14), and zein microspheres containing an entrapped protein polysaccharide, PS-K used for cancer immunotherapy (15).

In this study we investigated the encapsulation of essential oils into extremely small zein particles for use as controlled delivery systems of essential oils as antimicrobials which will minimize their interactions with other components found in feed formulations as well as the host.

Materials and Methods

Materials

Corn zein (decolorized) was obtained from Showa Sangyo Co. Ltd., Tokyo, Japan, distributed by Chugai Boyeki (America) Corp., New York, and zein F-4000 (not decolorized) was obtained from Freeman Industries, Tuckahoe, NY. Oregano and cassia 100% pure oils were obtained from Footeandjenks, Camden, NJ; red thyme 100% pure oil was from Frontier Aromatherapy, Norway, IA. Thymol and pepsin A (min 99.5%), EC 3.4.23.1 (1:10000), were from Sigma, St. Louis MO, and silicone fluid SF96/50 was from Thomas Scientific, Swedesboro, NJ. An Ultra Turrax T 25 high-speed dispersing apparatus was

178

manufactured by Jamke & Kunkel GMBH & Co. KG, Staufen, Germany, and a
reciprocal shaking bath (model 25) was from Precision Scientific, Winchester,
VA. A spectrophotometer UV-160 was from Shimadzu Corp., Kyoto, Japan.
Freeze-dried swine manure solids were generously donated by Dr. Patrick Hunt,
ARS, USDA, Florence, SC.

Encapsulation of Essential Oils in Zein Nanospheres

Powder particles containing essential oils were prepared according to the
method of Parris et al. (*9*) by dissolving 250 mg of oil and 1.0 g of zein in 15
mL of 85% ethanol. The solution was rapidly dispersed with high-speed mixing
into 40 mL of water containing 0.01% silicone fluid until a single phase was
formed (approximately 1 min). The opaque solution containing the encapsulated
oil particles was lyophilized overnight. The dry powder, which was loosely
attached to the lyophilizing bottle, was collected and stored in dessicators held at
0% relative humidity. Typical yields were between 65% and 75% of product.

Scanning Electron Microscopy

The topographical micro-and nanostructure of lyophilized powders made
from phase-separated zein and essential oil was examined by scanning electron
microscopy using a Quanta 200 FEG (FEI Co., Hillsboro, OR), operated in the
high vacuum, secondary electron imaging mode. Dry powders, with and
without essential oil, were deposited on a conductive carbon adhesive tab,
mounted on an aluminum specimen stub (Electron Microscopy Sciences,
Hatfield, PA). Excess, non-adherent powder particles were removed from the
surface of the adhesive tab with a jet of nitrogen at a pressure less than 80
pounds/square inch, then the surface was coated with a thin layer of gold by
sputter coating. Digital images were acquired at low magnification (50X) to
resolve the irregular shapes and sizes of powder particles and at high
magnification (50,000X) to resolve the shapes and size distributions of
constituent nanospheres.
Powder particles were dispersed in aqueous solutions of 24% ethanol and
negatively-stained with 2% uranyl acetate solution on carbon support films, air
dried, then examined by transmission electron microscopy at an instrumental
magnification of 45,000X.

Atomic Force Microscopy

The superficial topography of the powder particles was also compared by
atomic force microscopy, using a Nanoscope IIIa multimodal scanning probe

microscope with TESP cantilevers (Veeco Corp., Santa Barbara, CA) in the phase-shift mode. Dry powder particles were deposited on carbon adhesive tabs on metal disks (Ted Pella, Inc., Redding, CA) and micrometer areas were scanned at low set-point ratio settings for light, intermittent contact.

Laser Scanning Confocal Microscopy

The localization of the essential oil within the microstructure of powder particles was examined with the hydrophobic probe, Nile Red, Greenspan and Fowler (*15*), using a TCS-SP laser scanning optical microscope system (Leica Microsystems, Exton, PA) equipped with a 63X water immersion lens mounted in an IRBE optical microscope. Ten milligram aliquots of powders were suspended in one milliliter aliquots of distilled water. Fifty microliter volumes of suspended powder particles were transferred to glass-bottom dishes (MatTek Corp., Ashland MA), before and after addition of Nile Red (5 micrograms/mL). Samples were excited with the 488 nm line of an Argon laser. Autofluorescence and fluorescence spectra and fluorescence images from Nile Red were collected in the range of 500 to 690 nm and in two separate channels, 530-560 nm and 590-620 nm.

Transmission Electron Microscopy

The internal ultrastructure of embedded, thin sectioned powder particles, was examined with a CM12 scanning-transmission electron microscope (FEI, Co., Hillsboro, OR). About 10 milligram aliquots of dry powders were immersed in 10 milliliter volumes of 2.5% glutaldehyde in 0.1M imidazole buffer (pH 7.0) and stored in sealed vials before further processing. For embedding and thin sectioning, the suspended powder particles were collected by centrifugation and the resulting pellets were washed in imidazole buffer to remove glutaraldehyde, then reacted with 2% osmium tetroxide solution buffered with the imidazole buffer for 2 h, followed by dehydration in a graded series of ethanol solutions (50, 80 and 100%) for 1 h. The pellets were infiltrated with propylene oxide and a 1:1 mixture of propylene oxide and epoxy resin mixture, subsequently embedded in epoxy resin and cured for 48 hours at 55 °C. Thin sections of embedded powder pellets were cut with diamond knives, mounted on grids coated with carbon support films and stained with solutions of uranyl acetate and lead citrate. Photographic images were recorded at instrumental magnifications of 5,000X and 45,000X.

Determination of Encapsulated Oil

Oil entrapped in the zein particles was quantified after all the oil was removed by extraction of 10 mg of the lyophilized particles 3 times with 1 mL of ethyl acetate and comparing its absorbance to standard curves constructed at 275 nm for red thyme and oregano, at 276 nm for thymol, and at 281 nm for cassia oil. Zein was not soluble in ethyl acetate and did not interfere with the spectrophotometric determination.

Enzymatic Digestion of Zein Nanospheres

A modification of the methods described by and Liu et al. (*11*) and Johnson (*17*) was used for the in vitro digestion of zein particles. Powder zein particles (100 mg) and 10 mg of pepsin were suspended in a flask containing 100 mL of 0.01 M KH_2PO_4-citrate buffer + 0.5% Tween-20, pH 3.5. The flask was agitated at 50 rpm and 37 °C. Three milliliter samples were removed periodically. The reaction was quenched with 0.3 mL of 0.1 N NaOH, and the absorbance was measured at 280 nm.

Nonenzymatic in Vitro Release of Essential Oils from Zein Nanospheres

Powder particles containing encapsulated oil, 100 mg, were placed into a flask containing 100 mL of 80% phosphate buffer saline (pH 7.4) + 24% ethanol and agitated at 50 rpm and 37 °C. Samples (3 mL) were withdrawn periodically. Their absorbance was measured at the appropriate wavelength.

Gel Electrophoresis

SDS-PAGE of corn zein incubated in the presence of swine manure was carried out on a Phast System Pharmacia (Piscataway, NJ) with a phast gel of 20% acrylamide. Fifty milligrams of swine solids and 40 mg of zein were placed in five Eppendorf tubes. To each tube were added 0.5 mL of 10% SDS and 0.5 mL of 0.44M Tris buffer, pH 8, and then they were mixed. A total of 50 µL of 2-mercaptoethanol (2-ME) was added to the tubes after incubation at 37 °C for 0, 15, 30, 45, and 60 min and then heated at 100 °C for 10 min to stop digestion of zein. After cooling, the tubes were centrifuged at 14 000g for 4 min, the supernatant was passed through a 50 000 nominal molecular weight limit (NMWL) filter unit, Millipore Corporation (Bedford, MA), and the filtrate was analyzed. Gels were stained with 0.2% (w/v) Coomassie R350 dye. Molecular weight standards, Bio-Rad (Richmond, CA), and their corresponding molecular weights were as follows: phosphorylase *b* 97 000; bovine serum albumin (BSA), 66 200; ovalbumin, 42 699; carbonic anhydrase, 31 000; soybean trypsin inhibitor, 21 500; lysozyme 14 400.

Results and Discussion

Preparation of essential oil encapsulated zein nanospheres by phase separation was found to be more rapid and less tedious than solid-in-oil-in-water (S/O/W) emulsion (18) or chemical conjugation (10) methods previously reported. As described in Materials and Methods, essential oil and zein were dissolved in aqueous alcohol and rapidly mixed in water containing a small amount of dispersant. The opaque suspension was lyophilized overnight. The principal component in red thyme and oregano is thymol (λ_{max}=275 nm) and cinnamaldehyde (λ_{max}= 280 nm) in cassia oil. The oils were quantitatively extracted from the zein particles using ethyl acetate without interference from the decolorized zein protein, which was not soluble in ethyl acetate at the same concentration. Encapsulation efficiency of all oils (oil in nanospheres vs. oil added) was 87% except for cassia oil, which was 49%. Loading efficiency of all oils (oil in nanospheres vs. amount of nanospheres) was 22% except for cassia oil, which was 12%. These differences could be attributed to the greater solubility of cassia oil in 85% ethanol compared to the other oils.

Both scanning electron microscopy and atomic force microscopy of the powder particle surfaces reveal areas of close-packed spheres with a narrow range of diameters on a nanometer scale. In high vacuum-secondary electron images, at low magnification the particles were irregular in shape; the largest particles were around 50 micrometers wide, and the smallest were a few micrometers in the largest dimension (Figure 1A). At high magnification, the particle surfaces consisted of a layer of close packed nanospheres, ranging from less than 25 to over 200 nm in diameter. Except for differences in size-distribution between zein and zein-oregano oil, the nanospheres were similar in shape and packing (Figure 1B and 1C), and the superficial contours of individual nanospheres were smooth with few exceptions, where edges were contiguous. Analysis of the particle size distribution in several images indicated that the oregano oil encapsulated particle population had larger nanospheres than zein alone. We found that 31% of the oregano encapsulated nanospheres had mean diameters greater than 100 nm compared to 19% for the zein alone nanospheres. When the powder particles were dispersed into nanospheres by suspension in 24% aqueous ethanol solution (the final alcohol concentration of the preparation before lyophilizing) and negatively stained for bright field imaging by transmission electron microscopy, the nanospheres appeared to have a very regular circular profile with a uniform internal electron density, consistent with a spherical shape (Figure 1D). Phase contrast - atomic force microscopy in the intermittent contact mode of operation also revealed similar patterns of organization of the nanospheres (Figure 2A and 2B). No differences in particle contents was observed in phase-shift images of zein-oil samples, and images of zein were similar to those of zein-oil samples when compared over

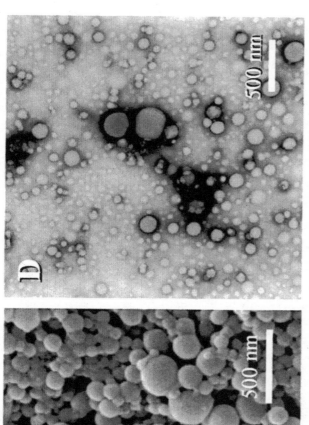

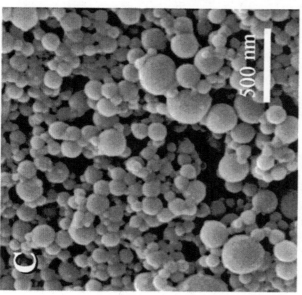

Figure 1. Scanning (A, B, C,) and transmission (D) electron microscope images of zein and zein-oregano oil samples. A. Particles of dry powder after lyophilization; B. Surface of zein powder particle illustrating nanospheres and C. Surface of zein-oregano oil powder particle illustrating nanospheres; D. Negatively stained bright field TEM image of zein-oregano oil nanospheres dispersed from suspension of powder particles in 24% aqueous ethanol solution.

184

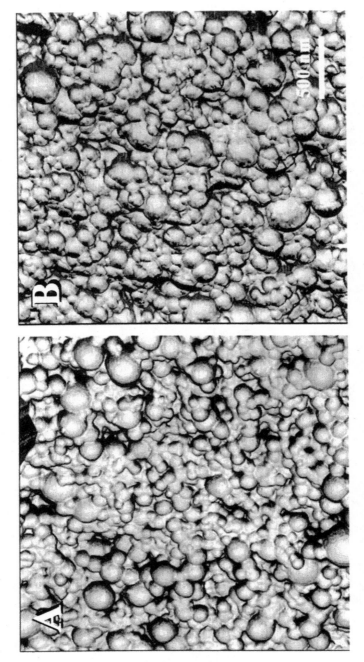

Figure 2. Phase-shift images, obtained by atomic force microscopy, of the surfaces of zein (A) and zein-oregano oil (B) powder particles illustrating comparable properties of the constituent nanospheres.

micrometer-sized areas. If the zein and oregano oil phases were separated and arranged or located in accessible, resolvable areas on the surfaces of the nanospheres, phase-shift images might be expected to detect the phase composition as differences in the degree of phase-shift in the cantilever oscillation (*19*). However, comparison of zein and zein-oregano oil particle surfaces at similar set-point ratios and other instrumental conditions indicated very similar patterns of variation in phase shift, ranging from around 1^0 to 2.5^0, over the surfaces of superficial nanospheres (Figure 2).

Efforts to localize oil in powder particles by fluorescence from Nile Red stain using confocal laser scanning microscopy were equivocal because the zein particles fluoresced at equal intensities as the zein-oregano oil powder particles over the range of emission expected for oil or lipid (*16*). Emission spectra of zein and the zein oil particles share a common maxima at 610 nm but the zein oil particles have an additional shoulder at 590 nm. Both exhibit autofluoresence from 500 to 650 nm when compared to the spectra for oregano (Figure 3A). The second derivative of the shoulder at 590 nm in the emission spectra for zein oil Nile Red particles is clearly resolved from zein Nile Red particles and oregano in oil indicating that the shoulder is clearly unique to the zein-oregano oil particles (Figure 3B). Efforts to differentiate between the zein and zein-oregano oil particle by confocal analysis, however, was unsuccessful (Figures 3A and 3B). This could be attributed to binding of a hydrophobic compound or compounds to zein. Momany et al. (*20*) have shown, through molecular model simulations, that the natural carotenoid, lutein fits into the core of the triple superhelix of 19 kDa (Z19) without perturbing its conformation significantly. In this study we used decolorized zein in order to avoid absorbance interferences of carotenoids with the essential oils. We have shown, after extraction with ethanol, that lutein and zeaxanthin, (the pigment of yellow corn), were present in zein F-4000 but not in the decolorized Showa zein used for this study and not responsible for the similarity between the zein and zein-oregano oil particles.

Attempts to stain the oil differentially with osmium tetroxide solutions on dispersed nanoparticles resulted in uniform electron density except for variations expected from particles with different diameters. Transmission electron microscopy of embedded, thin sectioned and stained samples illustrates the internal microstructure and nanostructure of zein and zein-oregano oil particles (Figure 4A and 4B). Surprisingly, the electron density of the zein matrix appeared homogeneous, with no obvious evidence of the expected nanospheres that were visualized in samples prepared for scanning and atomic force microscopy. The loss or dissolution of protein ultrastructure might be due to the effects of ethanol and propylene oxide that were used for infiltration and embedding, on the zein nanospheres, even though the protein was crosslinked with glutaraldehyde (Figure 4A). However, in addition to a homogenous (protein) matrix, the thin sections of zein-oregano oil particles contained numerous, uniformly distributed, electron-dense circles and rings (Figure 4B) with dimensions approximating the sizes of nanospheres seen in topographical images.

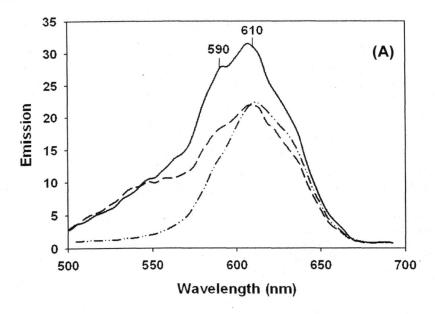

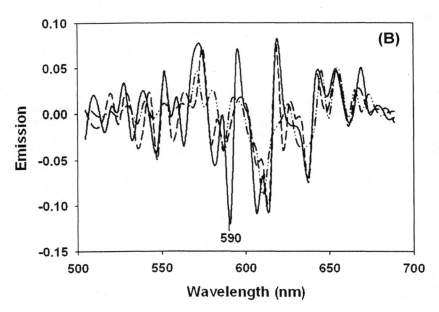

Figure 3. (A) Emission spectra of autofluorescence and fluorescence: oregano oil (—) zein nanospheres (---); oregano oil in zein nanospheres (—), stained with Nile Red. (B) Second derivative of emission spectra from Figure 3.

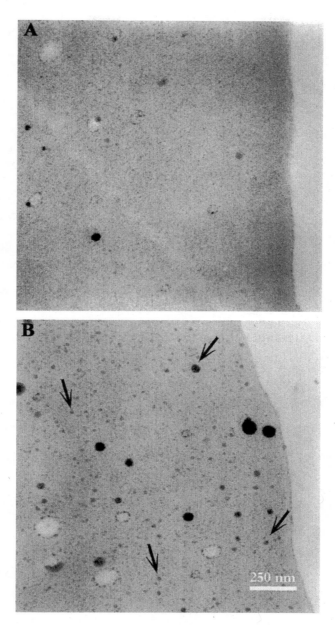

Figure 4. Thin sections of osmium tetraoxide-stained, plastic embedded particles of zein (A) and zein-oregano oil (B). In contrast to the relatively uniform electron density in (A), arrows in (B) indicate typical small circular and electron-dense rings.

Regardless of whether the oil is encapsulated or entrapped in the zein particle, it is important that the particle be stable under physiological conditions and that the oil be efficiently delivered to the desired site. The pH of the stomach may be as low as 1.0 and reach a pH between 3.0-4.0 due to the buffering capacity of food proteins (16). In this study the particles were subjected to in vitro digestion with pepsin (10:1) pH 3.5 at 37 °C. When zein samples were first introduced into the buffered solution, the particles formed aggregates that gradually dispersed and completely dissolved after 52 h (Figure 5). Less than half the particles were digested after 4 h, the typical gastric emptying time for humans and some monogastric animals. This indicates the possibility that zein particles will protect most of the essential oil from being released in the stomach. The oil will be released later in the small and large intestines.

Since bicarbonate ions are known to be secreted after the mixture enters the duodenum (15), it would be necessary to determine the effectiveness of the essential oils as an antimicrobial under neutral conditions. To illustrate the effectiveness of the essential oils as an antimicrobial agent, conditions in Figure 6 compare the nonenzymatic in vitro release of oregano from zein nanospheres in phosphate-buffered saline, pH 7.2, with and without ethanol. The ethanol content of the buffer was adjusted to 24% to minimize aggregation of the particles and to release the oregano oil more uniformly. After 24 h, in the presence of ethanol, 83% of the oil was released compared to 61% without alcohol. Little, if any, oregano appears to coat the particle surface since no initial burst of oregano was released which indicated that the oregano oil is encapsulated or entrapped within the particle. In the presence of alcohol, approximately 60% of the oil was released at a relatively constant rate for 4 h and at a slower rate for the next 20 h, after which no more oregano oil was released. Similar nonenzymatic release patterns in vitro were observed for red thyme, cassia oil, and thymol (Figure 7).

In the large intestine of animals, microbes may digest the zein and increase the rate of release of essential oils. We found that digestion of zein occurred within minutes, as indicated by gel elctrophoresis of zein incubated in the presence of swine manure solids at 37 °C (compare lanes 2 and 3, Figure 8). After 30 min, no zein was detected and more low molecular weight peptides were present (lanes 5 and 6). The rapid hydrolysis indicated that zein digestion was due to the presence of microbial enzymes in the freeze-dried swine manure solids and not from enzymes being produced during the assay by growing microbes. Experimental verification was carried out by heating the solids at 60 °C for 30 min (to denature the enzymes) or by autoclaving the solids (to inactivate the bacteria) before mixing with zein. In both cases the zein was not hydrolyzed which indicated that preformed microbial enzymes, in the freeze-dried manure, were responsible for hydrolysis of zein initially observed.

Encapsulation of essential oils in zein nanospheres by phase separation is a rapid and simple method. Particles appear to have limited digestibility in the

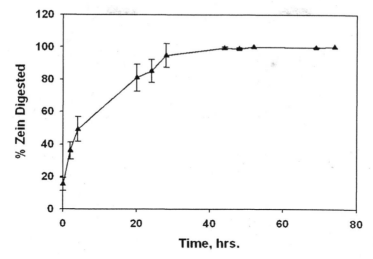

Figure 5. Enzymatic in vitro digestion of zein nanospheres with pepsin (10:1) in 0.01 M KH₂-citrate buffer + 0.5% Tween-20, pH 3.5 at 37 °C monitored at 280 nm. Data points are the mean of 3 replicates ± 1 standard deviation. (Reproduced with permission from 9. Copyright 2005 American Chemical Society.)

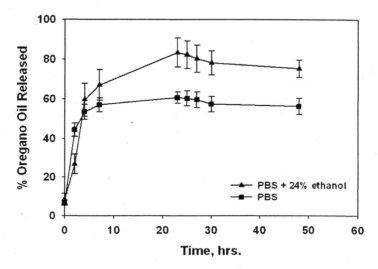

Figure 6. Nonenzymatic in vitro release of oregano oil from zein nanospheres in phosphate buffered saline (PBS) and PBS + 24% ethanol, pH 7.4, at 37 °C. Data points are the mean of 3 replicates ± 1 standard deviation. (Reproduced with permission from 9. Copyright 2005 American Chemical Society.)

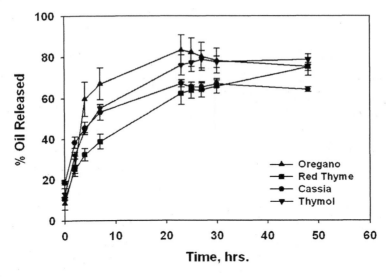

Figure 7. Nonenzymatic in vitro release of oregano (▲), red thyme (■), cassia oil (●), and thymol (▼) from zein nanospheres using conditions described for ▲ in Figure 8. Data points are the mean of 3 replicates ± 1 standard deviations. (Reproduced with permission from 9. Copyright 2005 American Chemical Society.)

stomach, slow release in the small intestine, and more rapid release in the large intestine. They can be used for oral or injectable administration of biological materials. In their encapsulated form, there should be little interaction of the essential oil with other components in the feed. Farmers and veterinarians should find application of the essential oils in this form to be easier to handle, less wasteful and hence more economical to use as a potential substitute for traditional antibiotics to prevent common intestinal infection in livestock. The in vivo release of essential oils in various animal species, their effect on microbial growth inhibition, and their probiotic effects will be studied before final formulations for feed applications are optimized.

Acknowledgment

We thank Dr. William Fett (deceased), John Minutolo, Guoping Bao, and Paul Pierlott for their technical assistance.

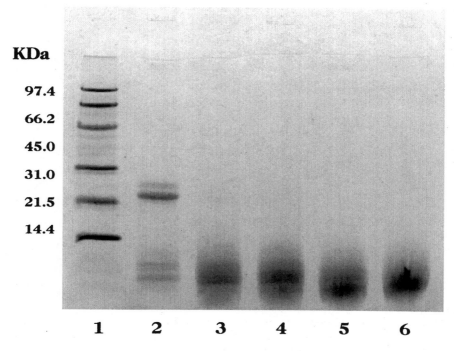

Figure 8. Digestion of zein by preformed microbial enzymes.
SDS-polyacrylamide gel electrophoresis of zein incubated with swine manure
solids at 37 °C: (lane 1) molecular weight standards; (lanes 2-6) zein-manure
samples after incubation for 0, 15, 30, 45, and 60 min, respectively.

References

1. Parry, J. W. In *Spices Their Morphology, Histology and Chemistry;* Parry, J. W., Eds.; Chemical Publishing Co. Inc.: New York, NY, 1962; pp. 187-189.
2. Cowan, M. M. *Clin. Microbiol. Rev.* **1999**, *12*, 564-582.
3. Bose, S. M.; BhimaRao, C. N.; Subramanyan, V. *J. Sci. Ind. Res.* **1949**, *8*, 157.
4. Bose, S. M.; BhimaRao, C. N.; Subramanyan, V. *J. Sci. Ind. Res.* **1950**, *9B*, 12.
5. Lambert, R. J. W.; Skandamis, P. N.; Coote, P. J.; Nychas, G. J. E. *J. Appl. Microbiol.* **2001**, *91*, 453-562.
6. Draughon, F. A. *Food Technology* **2004**, *58,* 20-28.

7. Langer, R.; Peppas, N. A. *AIChE Journal* **2003**, *49*, 2990-3006.
8. Peppas, N. A.; Bures, P.; Leobandung, W.; Ichikawa, H. *European Journal of Pharmaceutics and Biopharmaceutics* **2000**, *50*, 27-46.
9. Parris, N.; Cooke, P. H.; Hicks, K. B. *J. Agric. Food Chem.* **2005**, *53*, 4788-4792.
10. Hurtado-Lopez, P.; Murdan, S. *J. Pharmacy and Pharmacology* **2006**, *58*, 769-774.
11. Liu, X.; Sun, W.; Wang, H.; Zhang, L.; Wang, J. *Biomaterials* **2005**, *26*, 109-115.
12. Suzuki, T; Sato, E.; Matsuda, Y.; Tada, H.; Unno, K.; Kato, T. *Chem. Pharm. Bull.* **1989**, *37*, 1051-1054.
13. Demchak, R.J.; Dybas, R.A. Photostability of Abamectin/Zein Microspheres. *J. Agric. Food Chem.* **1997**, *45*, 260-262.
14. Tice, T. R.; Gilley, R. M. *Journal of Controlled Release* **1985**, *2*, 343-352.
15. Matsuda, Y.; Suzuki, T.; Sato, E.; Sato, M.; Kotzumi, S.; Unno, K.; Kato, T.; Nakai, K. *Chem. Pharm. Bull.* **1989**, *37*(3), 757-759.
16. Greenspan, P.; Fowler, S. *J. Lipid Res.* **1985**, *26*, 781-789.
17. Johnson, L. R. In *Gastrointestinal Physiology*, Johnson, L. R., Ed.; 6th ed.; Mosby, Inc.: St. Louis, MO, 2001.
18. Morita, T.; Sakamura, Y.; Horikiri, Y.; Suzuki, T.; Yoshino, H. *Journal of Controlled Release* **2000**, *69*, 435-444.
19. Scott, W. W.; Bhushan, B. *Ultramicroscopy* **2003**, *97*, 151-169.
20. Momany, F. A.; Sessa, D. J.; Lawton, J. W.; Selling, G. W.; Hamaker, S. A. H.; Willet, J. L. *J. Agric. Food Chem.* **2006**, *54*, 543-547.

Chapter 11

A Kinetic Study of Poorly Water Soluble Drug Released from Pectin Microcapsules Using Diffusion/Dissolution Model

Zayniddin Muhidinov[1,*], Jamshed Bobokalonov[1], LinShu Liu[2], and Reza Fassihi[3]

[1]Chemistry Institute of Tajikistan Academy of Sciences, 299/2 Ainy Street, 734063 Dushanbe, Tajikistan
[2]Eastern Regional Research Center, Agricultural Research Service, U.S. Department of Agriculture, 600 East Mermaid Lane, Wyndmoor, PA 19038
[3]School of Pharmacy, Temple University, 3307 North Broad Street, Philadelphia, PA 19140

A new microcapsular system was developed for the controlled drug delivery from pectins that obtained from various sources, with different molecular weight and the degree of esterification. The release kinetics of poor water soluble drug from the pectin microcapsules was investigated in simulated gastrointestinal fluids using prednisolone as a model drug. The work evaluated the effects of pectin macromolecular characteristics, the nature of surfactant and manufacturing conditions on the release kinetics of encapsulated drugs. The correlation between emulsion systems and drug release profiles was studied through the diffusion/dissolution number, which represents the combination of dissolution and diffusion kinetic parameters in one parameter. The microparticular diffusion coefficients determined by two different kinetics models are much smaller than analogues for other microparticular systems, indicating the critical step of intraparticular diffusion in drug released from pectin microcapsules. The highest value of drug dissolution/diffusion number was obtained for microcapsule from high methoxylated apple pectin in the presence of anionic surfactants and calcium ions, rather than for the systems of highly charged citrus pectin. Capsules prepared by the use of ethyl acetate also showed retarded drug release, however the amount of drug encapsulated was much less than those from other emulsion systems. All the microcapsules had drug release half-life time ($t_{50\%}$) of more then 5 hour. The rapid release of prednisolone was achieved in the presence of pectinase. These results suggested the application of biodegradable pectin polysaccharides in the production of vastly diverse drug carrier systems for colon-specific drug delivery.

The potential of calcium pectinate and zinc pectinate as drug carriers for colon-specific drug delivery has been evaluated *in vitro* and *in vivo* by the use of drug markers, both water soluble and water insoluble, either small organic compounds or active protein drugs (*1-7*). In these studies, calcium pectinate was formulated into films, gels, droplets, and most often, in the form of compressed tablets. The cross-linking of pectin with calcium ions inhibits the release of incorporated drug from the pectin tablets by suppressing the dissolution and swelling of pectin macromolecules. Nevertheless, it does not inhibit the diffusion of incorporated drugs from the surfaces of compressed tablets to the surrounding medium upon the drug's hydration. The release of drugs is affected by the swelling medium and drug diffusion. For tablets incorporated with water soluble drugs, the dissolution of drugs from the surfaces of the tablets is fast. As a result, water migrates into the matrix to replace dissolved drugs and create pores and channels. Fluid ingress promotes the extent and the rate of matrix swelling and creates a large surface area, which in turn, enhances the release of incorporated drugs at small are of the colon. To overcome this challenge, we have developed a new pectin-based drug delivery system in the presence of surfactants and counter-ions, which was more effective in both drug encapsulation and drug release in gastrointestinal (GI) tract (*8*).

Drug release from polymeric drug carriers is a complicated procedure, which coupling drug and solvent diffusion with drug dissolution and polymer erosion. Study on drug release pattern is of important in the development of controlled drug delivery systems. Drug release can be described in a simple form by recalling solution of Fickian diffusion equation with appropriate boundary condition (*9*). The new mathematical model considers drug dissolution, diffusion, polymer degradation/erosion, time-dependent system porosities and three-dimensional geometry devices using Monte Carlo Simulation (*10*). This model is able to describe the observed drug release kinetics accurately over the entire period of time, including 1) initial "burst" effect; 2) subsequent approximately zero order drug release phase; and 3) second rapid drug release phases. By this model important information, such as the evaluation of the drug concentration profiles within microparticles, can be calculated. There are some other modeling approaches, such as simplify the problem to purely erosion or diffusion controlled using the empirical and semiemperical mathematical model, such as Higuchi equation (*11*) and Baker (*12*).

In the present study, we discussed the release mechanisms of prednisolone, a model poor water soluble drug, encapsulated in pectin microcapsules using different diffusion and dissolution models.

Materials and Methods

Materials

Three types of pectin were used in this study. One type of pectin was prepared in house from apple pomace (AP) by membrane separation technology developed by Muhidinov (Dushanbe, Tajikistan). As molecular characterizations determined, the AP featured with: galacturonic acid content, 76 %; molecular weight (MW), 6-14 million; and the degree of esterification (DE), 43%. Other pectins were from citrus, GENU LW-12CG (MW >25 million; DE 35%) and Selendid 400 (MW 1.86 million; DE 5%), kindly provided by CP Kelco (Wilmington, Delaware, USA); they were recorded as CP12 and CS400, respectively. Prior to use for drug encapsulation, all pectins were purified by dissolving in deionized water at the concentration of 0.5 % (w/w) followed by centrifugation and filtration using an ultra filter (pore size, 0.45 μm).

Prednisolone, pectinase (EC 3.2.1.15), ethyl alcohol, ethyl acetate and calcium chloride anhydrous were purchased from Sigma-Aldrich (St. Louis, MO, USA). Sodium dodecyl sulfate (SDS), benzalkonium chloride (BzACl), and Span 80 were obtained from Amend CCI (Amend, NJ, USA). Hydrochloric acid, sodium hydroxide and potassium phosphate monobasic were provided from Fisher Scientific (Fairlawn, NJ, USA). Vegetable oil was obtained from market.

Microcapsule Preparation

Pectin microcapsules were prepared by emulsification-interface reaction technique consisting of two steps as described previously (*8*). Briefly, a primary oil-in-water microemulsion was prepared with the model drug, prednisolone, in the oil phase that was covered by an aqueous phase containing with either charged surfactants and pectin or non-charged surfactants and calcium chloride. To the primary emulsion containing with pectin, calcium chloride was added to initiate the cross-linking with pectin to form a pectin film capsules; to the primary emulsion carrying calcium, a pectin solution was added to form a second oil-in-water-in-water emulsion. Resultant pectin microcapsules were collected by ultra-centrifugation followed by washing with distilled water. For drug encapsulation efficiency, the supernatant obtained from centrifugation was combined with the washing solutions and filtered through a membrane filter (pore size, 0.45 μm; Millipore, Bedford, MA, USA). An aliquot of the filtrate was taken and diluted to an appropriate concentration, measured at 248 nm for the amount of non-encapsulated prednisolone using an UV-Vis spectro-photometer (Agilent 8453, USA). The amount of encapsulated drug was then calculated by subtracting the amount of prednisolone measured from the amount of prednisolone added. Alternatively, the encapsulated prednisolone was obtained by microcapsule digestion with pectinase followed by measuring the optical density of the digestion medium at the same wavelength. The mean results were expressed as (A) loading capacity: the percentage of drug in 100 mg dried material; and (B) encapsulation yield: the percentage of drug encapsulated.

Release Kinetics

Prednisolone Release from Pectin Microcapsules

The release kinetics of prednisolone from pectin microcapsules were investigated *in vitro* under conditions mimicking the GI tract: 0.02M HCl/KCl (pH 1.5), 0.02M NaOH/KH$_2$PO$_4$ (pH 6.4), and 0.02 M phosphate buffer containing pectinase at around 37°C. Microcapsules (30-100mg) were placed in a glass vial containing 60-200ml of dissolution medium. The vials with the contents were shaken in a water bath at 37°C. At desired time intervals, an aliquot of the release medium, about 1-4 ml, was taken out by a syringe equipped with a submicron filter. After measuring the O.D. at 248 nm for prednisolone content, the medium was placed back to the vial to keep the volume constant.

Each test was carried out in triplicate. Experimental data was analyzed against three different mathematic models and the dissolution/diffusion numbers (Di) of prednisolone release was obtained.

Evaluation of Release Kinetics

The mechanism applied for the evaluation of prednisolone release was the Higuchi model (*13*), which was developed for diffusive release of drug from porous matrix when the drug is present in excess of its solubility. In this pseudo steady-state model we assume that the dissolution of the drug does not influence the release rate. In other word, drug release is diffusion control with dissolution being relatively rapid. For a system with finite sink where the bulk solution concentration is positive, a kinetic equation can be expressed as (15, 17):

$$f_t = M_t/M_0 = 3K_r t^{1/2} - 3(K_r t^{1/2})^2 + (K_r t^{1/2})^3 \tag{1}$$

where, f_t is the fraction of the material released at time t, K_r is the release rate constant.

This expression can be written in a linear form as:

$$(1 - f_t)^{1/3} = 1 - K_r t^{1/2} \tag{2}$$

By plotting the left-hand side of the above expression as a function of the square root of time, a linear plot with slope K_r is obtained.

The release of a poor water soluble drug from microcapsules includes four steps (16): Step 1, diffusion of media solution into microcapsule; step 2, dissolution of drug in microcapsules; step 3, transfer of the dissolved molecule from the core of capsules to the boundary film; and the final step, transport of dissolved molecules from boundary film to the surface of the microcapsules.

Since steps 1 and 3 are osmotic pressure-dependant, it can be neglected. The second and the fourth step are generally contributed to dissolution of incorporated drugs and the diffusion of the drug dissolved. In most cases, the intra-particular diffusion is the slowest, rate-limiting step of the sorption process. This process may be fitted in to mathematical model of Baker-Lonsdale (12) developed for the diffusion controlled systems.

Another model to estimate the intra-particular diffusion was described by Urano and Tachikawa (15). The mass transfer for intra-particular diffusion defined by Nakai-Tachikawa model is governed by the differential equation below:

$$\frac{\partial q}{\partial t} = D_i'(\frac{\partial^2 q}{\partial r} + \frac{2}{r}\frac{\partial q}{\partial r}), \tag{3}$$

where, r is the radial variable, q is the sorption capacity, and D_i' is the global diffusion coefficient. The solution of this differential equation, if the adsorption rate is independent of the stirring speed and external mass transfer is not the limiting step of the sorption, is given by the following equation:

$$f\left(\frac{q_t}{q_m}\right) = -\left[\log\left(1 - \left(\frac{q_t}{q_m}\right)^2\right)\right] = \frac{4\pi^2 D_i' t}{2.3d^2} \tag{4}$$

where q_t is the concentration of the drug in the particles at time t, q_m is the concentration of the drug in the particles at equilibrium $(t \rightarrow \infty)$, and d is the mean particle diameter. Plotting $f(q_t/q_m)$ versus time allows one to determine the global diffusion coefficient if intra-particular diffusion is the limiting step of the drug release process.

Diffusion/Dissolution Model

Drug dissolution cannot be the only mechanism by which a drug is transported out of microcapsules. This transport occurred by subsequent drug diffusion. Hence, a common approach to characterization of dissolution effect is to expand on the diffusion equation. The diffusion/dissolution model incorporates a linear first –order dissolution term into Fickian diffusion equation (11).

$$\frac{\partial C}{\partial t} = D\left(\frac{\partial^2 C}{\partial r^2} + \frac{2}{r}\frac{\partial C}{\partial r}\right) + k\left(\in C_{sat} - C\right) \tag{5}$$

In this equation, k represents a first order dissolution constant (time $^{-1}$), $\in$ is the porosity of microcapsules, and C_{sat} denotes the saturated concentration of the drug in the system. Hence $\in C_{sat}$ represent the equivalent drug saturation concentration in the pores. The notations for the parameters C is drug concentration at time t and position r from the center of the capsules, D is effective diffusion coefficients of the drug in the polymer capsules.

Equation (3) can be transferred into a dimensionless form by defining the following dimensionless parameters:

$$\xi = r/R \tag{6}$$

$$\tau = Dt/R \tag{7}$$

$$\psi = 1 - C/(\in C_{sat}) \tag{8}$$

where ξ is dimensionless radial position, τ is the dimensionless Fourier time, and ψ is the dimensionless concentration. A new dimensionless number defined by Harland and co-workers is Di, which they named the dissolution/diffusion number.

$$Di = kR^2/D \tag{9}$$

This number expresses the relative importance of the dissolution and diffusion terms in relation to the overall release process.

The goodness of fit of the model to the experimental data is measured by an error term determined by the sum of the squared errors of the cumulative mass fractional drug released at each time point.

Results and Discussion

Microcapsule Characterization

The characteristics of typical pectin microcapsules prepared in this study are shown in Table I. In general, all microcapsules had relatively small particle size (< 4 µm), large particle number (No./ml ≥ 10^6), and high surface areas. However, all these parameters varied with manufacturing conditions, the types of surfactant and the types of pectin in use. The largest particle size of 3.72 µm was obtained for sample [#]16, which was prepared from low D.E. citrus pectin. The largest particle number (No./ml) was obtained for sample [#]11, which was prepared from the high D.E. citrus pectin that had the highest molecular weigh. Although all samples were found to bear a similar particle size distribution (data not shown) and were stable under experimental conditions, the size distribution

Table I. Physicochemical Characterization of Microparticles

Sample Code	Emulsion and Components	Total Particle No. per ml	Mean Diameter, µm	Specific Surface Area S_{sp}^{a}, cm^{-1}	%, Yield of Encap- sulation
6	O/W, VO- AP-SDS07-Ca^{2+}	14 276 456	1.65	6060.61	84.5
9	O/W, VO-AP-SDS05-Ca^{2+}	12 356 457	1.83	5464.48	18.2
10	O/W, VO-AP-BzACl	8 765 321	1.92	5208.33	14.6
11	W/O, CP12-VO-Span80	25 198 300	1.85	5405.41	75.0
12	O/W, VO-AP-BzACl	9 335 780	2.95	3389.83	28.6
13	O/W, VO-AP-BzACl-Span80	13 916 520	2.98	3355.70	40.2
14	O/W, EtAc-AP-BzACL	5 980 960	2.10	4761.90	31.0
16	O/W, EtAc-CS400-BzALCl	1 866 640	3.72	2688.17	22.7

[a]S_{sp}- specific surface area is ratio of surface area and volume of particles (S_{sp}=S/V).

of samples [#]12, [#]14 and [#]16 were clearly bi-modal in 2 days (8). The increase in the ratio of O/W and pectin concentration in the water phase from 1.0% (samples [#]10) to 1.5% (sample [#]12) showed no effect on the number of particles harvested, but it resulted in a slightly increase in particle size and a remarkable decrease in specific surface area, and also suppressed particle aggregation. Consequently, the increase of O/W ratios and pectin concentration enhanced drug incorporation. The addition of nonionic surfactant, Span 80, enhanced the yield of drug encapsulation (comparing samples [#]12 and [#]13). The replacement of vegetable oil with ethyl acetate did not cause statistically significant increase in drug loading (comparing samples [#]12 and [#]14), but required an additional step to remove organic residuals. Furthermore, casual observation in a three-month period revealed that the sample [#]13 had the highest stability.

Table I also indicates that the formulations with higher particle number and smaller mean particle diameter, for instance, samplers [#]6 and [#]11, have higher encapsulation efficiencies than other formulations. In addition, the increase in SDS surfactant concentration would lead to the increase of drug incorporation (comparing samples [#]6 and [#]9). Presumably, the increase of SDS concentration resulted in highly charged primary emulsion and an increase in the total number of microcapsules, which, in turn, enhanced drug encapsulation. These results indicated that drug loading efficiency could be improved by altering formulation conditions.

Release of Prednisolone from Pectin Microcapsules

In Figure 1, we compared the release kinetics of prednisolone from two formulations, samples [#]6 and [#]9, which prepared under same conditions, except for the concentrations of SDS surfactant. With a higher surfactant concentration, the microcapsules have higher particle number and larger surface areas (Table I), thus, it is inferable that the microcapsules should have a thinner wall than those prepared with lower surfactant concentration. A thinner membrane is less difficult for prednisolone to diffuse through into a release medium than a thicker one. This was confirmed by Figure 1, drug release from sample #6 at a higher rate than from sample [#]9. Figure 1 also showed that the release of prednisolone depended on the pH values of dissolution media. Drug released into low pH medium at a faster rate than that into a solution at neutral pH. Furthermore, the addition of pectinase enhanced drug release, because it caused the degradation of microcapsules.

The characteristics of prednisolone released from other 4 formulations (samples [#]10-13) into low pH medium were shown in Figure 2. Sample [#]12 showed the lowest release rate for an incubation period of 30 h. This could be attributed to its lower total particle number, larger particle size and the lower total specific surface areas. The release profile of prednisolone from sample #16 is similar (data not shown). However, if a non-charged surfactant, Span80, was used in creating a primary emulsion system, an increase in the release rate of

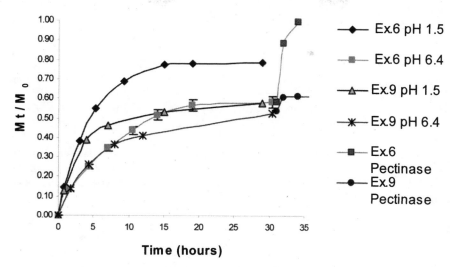

Figure 1. Prednisolone release rate from microparticulates Ex. 6 and Ex. 9 in HCL/KCL buffer (USP, pH 1.5), NaOH/KH₂PO₄ buffer (USP, 6.4) and subsequent Pectinase EC 3.2.1.15 (pH 6.07) treatment.

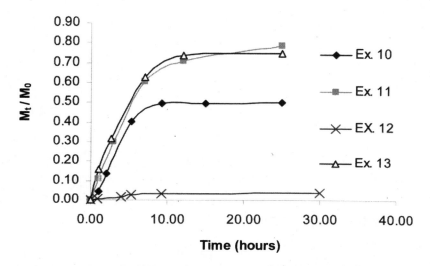

Figure 2. Rate of Prednisolone release from Microparticulates 10-13 at medium pH 1.5.

incorporated drugs was observed regardless the values of the total particle number, particle size and the specific surface areas (Sample [#]11 and [#]13). The effect of particle size on drug release rate also can be seen by comparing sample [#]10 with sample [#]12. Although both the samples were formulated under same conditions, sample [#]12 was from a higher concentrated pectin solution, thus had a higher particle size; sample [#]10 displayed a higher drug release rate than sample [#]12.

Figure 3 showed the release profiles of prednisolone from samples [#]10-13 into neutral pH medium containing with or without pecrtinase. Results similar to that shown in Figures 1 and 2 were obtained

These results (Figures 1-3) demonstrated that the prednisolone release kinetics from pectin-derived microcapsules could be controlled by altering the formulation conditions, the MW and type of pectin, and the type and amount of surfactant.

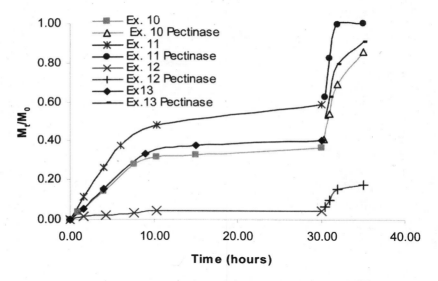

Figure 3. Rate of Prednisolone release from Microparticulates 10-13 at medium pH 6.4 and subsequent pectinase (EC 3.2.1.15, pH 6.07) treatment.

Analysis of Drug Release Model

Higuchi plot, $(1 - f_t)^{1/3}$ vs. $t^{1/2}$, for formulations [#]10-13 was shown in Figure 4. From the linear plot, the release constant, K_r, of prednisolone released from pectin microcapsules at pH 1.5 or pH 6.4 was obtained. Correlated Dissolution/Diffusion Numbers, Di, and r^2 were calculated according to two mathematics models (*14, 15*) using the liner regression. They are shown in Table II.

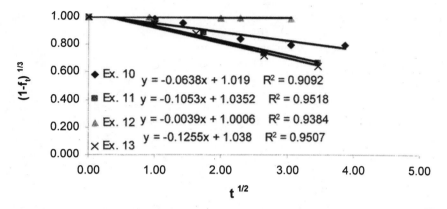

Figure 4. Higuchi plot for microcapsules of Exs. 10-13 at pH 1.5 Mean diameter of microparticles is 1, 92, 1.85, 2.95 and 2.98 mu respectively.

Table II. Mathematics Model Studied Diffusion/Dissolution for Drug Release from Microparticles in HCl/KCl (pH 1.5) and NaOH/KH$_2$PO$_4$ (pH 6.5) buffers media at 37°C.

Micro-particle code	Release rate (11) K_r (10^{-2}), h^{-1}				Diffusion (12) D (10^{-12}), cm^2/min				Diffusion (15) D (10^{-12}), cm^2/min			
	pH 1.5	r^2	pH 6.4	r^2	pH 1.5	r^2	pH 6.4	r^2	pH 1.5	r^2	pH 6.4	r^2
6	10.08	0.98	4.85	0.91	0.27	0.85	0.46	0.95	0.21	0.87	0.33	0.99
9	5.87	0.98	4.24	0.99	0.31	0.94	0.56	0.93	0.17	0.96	0.25	0.97
10	6.38	0.91	2.79	0.82	7.10	0.92	4.34	0.94	0.96	0.92	0.37	0.87
11	10.53	0.95	6.28	0.95	1.35	0.84	1.27	0.90	0.91	0.92	0.31	0.90
12	0.39	0.94	0.43	0.99	4.54	0.97	4.88	0.76	0.14	0.93	0.13	0.99
13	12.55	0.95	8.57	0.84	3.53	0.82	3.04	0.81	2.10	0.85	0.86	0.85
14	1.37	0.96	0.54	0.89	1.28	0.95	2.34	0.96	0.22	1.00	0.29	1.00
16	0.89	0.75	1.70	1.00	1.89	0.84	9.82	0.98	0.29	0.90	0.43	1.00

Assuming that drug dissolution and the diffusion of dissolved drug contribute to the kinetics of drug release (Figures 1-3), the drug release profile was investigated by fitting the drug release data into Higuchi (*11*) model (Figures 4) and Urano and Tachikava (*15*) models (Figure 5). For the incubation period of 2-20 hours, data of drug release for all the samples were well fitted in a linear plot (r^2 =0.973- 0.999) of both Higuchi and Tachikava models. The release constant (K_r, h^{-1}) and diffusion constant (D, cm^{-2}) were can be obtained from the slopes. The dissolution and diffusion constant was reported as a dissolution/diffusion number (*13*), expressing the relative importance of the dissolution and diffusion terms in relation to the overall release process.

As discussed in the previous section, the release profile of prednisolone can be controlled by altering manufacturing conditions and the physical characteristics of resultant delivery vehicles, such as wall thickness and the particle size of the microcapsules, and the conformation changes of pectin macromolecules.

A higher concentration of prednisolone incorporated in microcapsules can result in a large concentration gradient across the capsule wall and, as a consequence, cause a high osmotic pressure within the capsules and accelerated the diffusion of prednisolone molecules through the capsule wall. At a solution pH higher than 3.5, pectin molecules in the walls of microcapsules are more swelled and the hydrated polymer chains increase wall density, which could retard drug diffusion. This was clearly seen in Figure 1 by comparing the two sets of plots (pH 1.5 *vs.* pH 6.4).

Figure 4 presents Higuchi plot for sample #10-#13. As expected, the release rate of the prednisolone decreased as increase in the wall thickness of capsules at high pH values (Table II). As film encapsulation method was changed using cationic BzA-Cl surfactant without of Ca^{++}, the release properties was improved. Sample #10, regardless of its low encapsulation efficiency, showed a good example for sustained drug release model. The slower release rates of the thicker capsules were obviously due to the decreased permeability of their shells. The retarded drug release from sample #12 could be attributed to the aggregation of microcapsules with increasing of pectin concentration, while increasing of oil content at emulsification step to favor produce more stable capsules. Visually, sample #13 was promising in form and physicochemical properties, but drug release parameters from this sample was higher then another.

Table III summarizes the major result in this work: the application of microcapsules, as a drug carrier. It combine the influence of physicochemical properties of studied emulsion systems and drug release kinetics parameters, such as 50% release time of drug release ($t_{1/2}$), and the diffusion/dissolution number (D_i). The half release times ($t_{1/2}$) for samples #9, #10, #14, and #16 is higher, both at pH 1.5 and at pH 6.4, and to range from 10 to 77 hours. They are roughly proportional to the wall thickness and osmotic pressure produced by drug encapsulated. This suggests that the increase of the film thickness has significant effect on the capsule wall structure and the apparent diffusion coefficient.

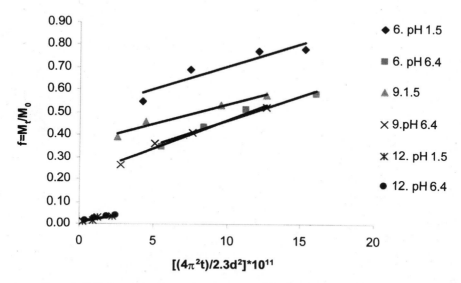

Figure 5. N-T model for microparticles 6-9 at medium pH 1.5 and 6.4.

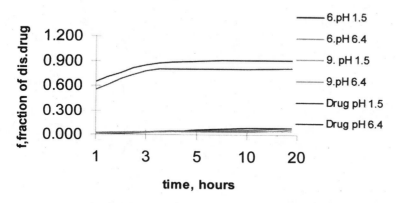

Figure 6. Drug dissolution in pH 1,5 and 6,4 and its contribution to the drug release from microcapsules

Table III. Drug Release Kinetics Parameters

| | Kinetics Parameters | | | |
| | $T_{50\%}$ h | | $Di=kR^2/D$ | |
Microparticle code	pH 1.5	pH 6.4	pH 1.5	pH 6.4
6	6.90	14.43	3553	1049
9	11.80	16.34	2824	1409
10	10.86	24.84	612	695
11	6.60	11.03	994	1733
12	177.70	161.16	602	709
13	5.55	8.09	1327	2205
14	51.70	128.3	687	205
16	77.86	40.76	1062	1368

Transport studies could be particularly valuable for understanding the realize kinetics of the encapsulated substances. A linear Nakai and Tachikava relationship (Figure 5) confirms that intraparticular diffusion is the limiting step of the drug release from the pectin microparticles.

The microcapsular system presented in this study had a significant advantage over those prepared from very concentrated solution of pectin via calcium cross-linking (4). This may be attributed to the fine regulated gelation process by pectin macromolecules to form thin film, rather than spontaneous formation of gel. The fact that the single emulsion-interface reaction process led to the production of intact microcapsules from every formulation listed in Table I, except samples #12 and #16, is consistent with principal of polymer precipitation. Thus, higher molecular weight polymers precipitate more readily than low molecular weight polymers, and, at a given dispersed: continuous phase ratio, complete precipitation is more easily achieved from dilute, rather then from concentrated, solutions.

Finally, we should comment on the release mechanism and the application of such microcapsules in controlled drug release. The release of pectin encapsulated drugs involves two processes: (i) the bulk solution diffuses into the capsules to extract the drug from oil phase, and (ii) the dissolved drug molecules diffuse out of the capsules. As we can see from Figures 6, the bare prednisolone crystals dissolve much faster than the encapsulated one. It is expected that at the early release stage, the prednisolone concentration within the capsules is very high, presumably close to its saturation solubility in the bulk solution. The same situation will remain until the drug in oil phase within each capsule exhausts. During this period, the permeability of the capsule wall controls the release rate; therefore, the drug release should be governed simultaneously by the saturation solubility of the drug and the permeability of the pectin capsule. This could explain why the release rate increases as the saturation solubility of the encapsulated drug increases, but proportionally decreases as the capsule thickness increases. In controlled drug release, a major problem related with microcapsules is the prominent initial burst and incomplete unloading of encapsulated drugs (*16, 17*). In the former case, a large portion of the drug was lost in a short period, which may cause acute toxicity and failure of controlled release. In the latter case, a portion of drug will remain within the carriers without release, which may affect the efficiency of the encapsulated drugs. In the case of low water-soluble cores, both the initial burst and the incomplete unloading of the encapsulated drug do not present a problem for pectin microcapsule systems. A complete release can always be achieved since the capsule walls are very thin. In this regard, the drug loading of PEM microcapsules is very efficient.

Conclusion

Pectin microcapsules sized between 1.65 and 3.72 μm have been prepared with calcium ion and different types of surfactant by O/W and W/O emulsion systems. The fact that the single emulsion-interface reaction process led to the production of intact microcapsules from every formulation listed in Table I, except for samples #12 and #16, is consistent with principal of polymer precipitation. Thus, higher molecular weight polymers precipitate more readily than low molecular weight polymers, and, at a given dispersed: continuous phase ratio, complete precipitation is more easily achieved from dilute, rather then from concentrated, solutions.

Correlation between physicochemical and drug release kinetics parameters was investigated. The effect of pectin macromolecular structure and the nature of surfactant on the manufacture of microcapsules were studied. The influence of the physicochemical properties of emulsion systems on drug release kinetics was evaluated through the diffusion/dissolution number, which represent combination of dissolution and diffusion kinetics parameter in one parameter. The release rate of the materials encapsulated in the pectin microcapsules was shown to be controlled by the abilities of different pectins to form a variety of microcapsule walls by taking advantage of the conformational flexibility of pectin polysaccharides in presence of different counterions. A linear Nakai and Tachikava relationship confirms that intraparticular diffusion is the limiting step of the drug release from the pectin micropaticles. The microparticular diffusion coefficients determined by two different kinetics models were much smaller, in comparison with analogues numbers for others microparticular systems, indicating that the diffusion process is much lower in pectin microcapsules. The highest value for drug dissolution/diffusion number was observed for the lower methoxylated apple pectin microcapsule formed from anionic surfactant and calcium ions rather than for the system of highly charged citrus pectin.

The higher superficial area and the slower *in vitro* degradation rate of the microcapsules fabricated in the present study suggest the potential uses as delivery systems for the controlled release of drugs and other active materials. Our results suggest that pectin capsules can be used to encapsulate low water-soluble drugs for controlled drug release.

References

1. Ashford, M.; Fell, J.; Attwood, D.; Sharma, H.; Woodhead, P. *J. Control. Rel.* **1993**, *26*, 213-220.
2. Sriamornsak, P.; Nunthanid, J. *J Microencapsulation* **1999**, *16*, 303-313.
3. Rubinstein, A.; Radai, R. *Eur. J. Pharm. Biopharm.* **1995**, *41*(5), 291-295.
4. Wong, T. W.; Chan, L. W.; Lee, H. Y.; and Heng, P.W.S. *J. Microencapsulation* **2002**, *19(4)*, 511-522.
5. Vandamme, Th. F.; Lenourry, A.; Charrueau, C.; Chaumeil, J. C. *Carbohydrate Polymers* **2002**, *48*, 219-223.
6. Liu, L.S.; Fishman, M. L.; Kost, J.; Hicks, K. B. *Biomaterials* **2003**, *24*, 3333-3343.
7. Pillay, V.; Danckwert, M. P.; Muhidinov, Z. K.; Fassihi, R. *Drug Dev. & Ind. Pharm.* **2005**, *31*, 191-207.
8. Muhidinov, Z. K.; Khalikov, D. Kh. Speaker, T.; Fassihi R. J. Microencapsulation **2004**, *21*(7), 729-741.
9. Crank, J. In *The Mathematics of Diffusion*, 2nd Ed.; Clarendon Press: Oxford, UK, **1975**, 89-103.
10. Siepmann, J.; Faisant, N.; Benoit, J-P. *Pharm. Res.* **2002**, *19*(12), 1885-1893.
11. Higuchi, T. *J. Pharm. Sci.* **1963**, *53*, 1145-1149.
12. Baker, R. W. In *Controlled Release of Biologically Active Agents;* Baker, R. W. Ed.; New York: John Wiley &Sons, 39-83.
13. Harland, S. H.; Pubernet, C.; Benoif, J. P.; Pappes, N. A. *J. Control. Rel.* **1988**, *7*, 207-215.
14. Cobby, J.; Mayerson, M.; Walker, G. C. *J. Pharm. Sci.* **1974**, *63*, 725-737
15. Urano, K.; Tachikawa, H. *Ind. & Eng. Chem. Res.* **1991**, *30*, 1897-1899.
16. Vayssieres, L.; Chaneac, C.; Tronc, E.; Jolivet, J. P. *J. Colloid Interface Sci.* **1998**, *205*, 205.
17. Herzfeldt, C. G.; Kummel, R. *Drug Dev. Ind. Pharm.* **1983**, *9*, 767.

Chapter 12

Protein Microspheres from Corn as a Sustained Drug Delivery System

Jin-Ye Wang[1,2*], Hua-Jie Wang[2], and Xin-Ming Liu[1]

[1]Shanghai Institute of Organic Chemistry, Chinese Academy of Sciences, 354 Fenglin Road, Shanghai 200032, China
[2]School of Life Science and Biotechnology, Shanghai Jiaotong University, 1954 Huashan Road, Shanghai 200030, China

Zein, as a sustained release material, was developed that permits continuous release of biologically active molecules from the microspheres, films and tablets. Ivermectin and heparin were chosen as the fat-soluble and the water-soluble model drugs, respectively. The heparin-loaded film or the ivermectin-loaded tablet were prepared based on drug-loaded zein microspheres that were obtained by the phase separation method and characterized by a scanning electron microscope and laser light scattering particle size analyzer. The biocompatibility of zein was evaluated, including human umbilical veins endothelial cells compatibility and hemocompatibility. Then the release in vitro and activity of drugs from the microspheres, films and tablets were determined. The results showed that zein and its degraded product had better biocompatibility; the sizes of microspheres could be controlled in 10nm-10µm and affected by the preparation conditions, such as concentrations of zein and drugs. Drugs in microspheres, films and tablets could be released slowly, in the mechanism of the diffusion of drugs through the matrix. Moreover, the drugs still kept the high activity.

The field of drug delivery system (DDS) is advancing rapidly and had been used widely, which involved such diverse fields as the pharmaceutical, agricultural, cosmetic and food industries and so on. For a drug delivery system application, slow release systems are of general interest since the lower frequency of administration leads a cut of labor and costs. It thus becomes possible to increase drug efficacy and decrease systemic adverse reactions compared to the traditional formulation that usually involves a mixing of the free drug powder with fodder (1). Carrier technology offers an intelligent approach for drug delivery by loading the drug in or on a carrier particle such as microspheres, nanoparticles, lipospheres, cylinders, discs, and fibers that modulates the release and adsorption characteristics of the drug (2).

Natural materials were thought to be biocompatible that means less thrombogenic and inflammatory, involving mainly polysaccharides and proteins, such as chitosan (3), starch (4), hyaluronic acid (5), albumin (6), casein (7), fibroin protein (8), silk protein (9). However, many problems for natural materials still existed, such as low encapsulation efficiency and drug loading, toxicity of residual solvent, stability of drugs, how to release the zero-order release and the target of the drug, etc. (10).

Zein is a native protein from corn and has shown its potential application as a biomaterial in our previous studies (11-13). In this chapter, we report the preparation and characterization of three different biodegradable drug-delivery formations based on zein. A series of drug-zein microspheres are prepared by phase separation method. Physical and morphological studies are carried out to examine the drug-protein microspheres by scanning electron microscope and laser light scattering particle size analyzer. Biocompatibility of zein and its degraded product were evaluated by 3-(4,5-Dimethylthiazol-2-yl)-2,5-diphenyl tetrazolium bromide method (MTT) and thrombin time assays. The releases in vitro of the ivermectin and heparin from the microspheres, film and tablet with or without the action of the enzyme were evaluated.

Materials and Methods

Materials

Zein with biochemical purity was purchased from Wako Pure Chemical Industries, Ltd. (Osaka, Japan), endothelial cells growth supplement (ECGS) was purchased from Upstate Biotechnologies Inc. (Lake Placid, NY, USA), heparin and thrombin were from Sigma (St Louis, MO, USA), ivermectin (IVM) was from Tongren Drug Company (Shanghai, China), and other reagents were reagent grade.

Cells Compatibility of Zein and its Degraded Product

Zein was dissolved in aqueous mixture of ethanol, and the ethanol content was decreased to 40% immediately to form zein microspheres suspended solution, then the film was prepared through volatilization at 37°C. The film was sterilized by UV for at least one hour and immersed into the RPMI 1640 nutrient fluid overnight before use. Zein degraded product was obtained from zein powder digested with pepsin (zein: pepsin=10: 1 w/w) in citric acid-NaH_2PO_4 (pH 2.2) at 37°C. The supernatant was freeze-dried to get the degraded powder for use.

Human umbilical veins endothelial cells (HUVECs) were harvested from human umbilical veins as described (14) and cells at passage 3 were digested by incubation with 0.025% trypsin/0.002% EDTA, resuspended with RPMI 1640 medium (20% FBS, 100 units/ml penicillin, 100 µg/ml streptomycin, 30 µg/ml ECGS and 90 µg/ml heparin) and seeded at $2.4×10^4$ cells/ml onto the surface of 96-well Corning culture microplate coated with zein film, other groups were zein degraded product adding groups (1 mg/ml) and culture microplate control group. After 2 hours, the culture medium in each well was taken out and the new medium (without or with zein degraded product) was added. The Corning microplate was as the control. The medium was refreshed every one day. HUVECs were photographed with inverted optical microscope everyday. At a shorter culture time, the non-adhered cells in the medium were counted with a hemocytometer to evaluate the adhesion of HUVECs under different conditions. 3-(4,5-Dimethylthiazol -2-yl)-2,5-diphenyl tetrazolium bromide (MTT) method was used to assess the proliferation of HUVECs after 2 hours, 1 days, 3 days and 5 days of culture. The attached cells on different substrates were fixed with 2.5% glutaraldehyde in PBS for at least two hours, washed with ultrapure Mili-Q water (PALL, USA) and freeze-dried. The samples were mounted on stubs and coated in vacuum with gold. Then cells were examined with a scanning electron microscope (SEM S-450, Hitachi, Japan).

Microspheres Preparation

Ivermectin Zein Microspheres

IVM-loaded microspheres were prepared using a phase separation procedure. For example, zein and IVM were dissolved in 66.7% ethanol. Then, ultrapure Mili-Q water was immediately added with vigorous mixing with an agitator (IKAMAG RCT basic, IKA, Germany) set at 9 at room temperature, and then the microspheres were lyophilized overnight before use.

The zein microspheres were mounted on brass stubs and sputter-coated with gold in an argon atmosphere prior to examination under SEM for its morphology. The mean diameter and particle size distribution of the microspheres were measured using a laser light scattering particle size analyzer (Zetasizer 3000HS, MALVERN, United Kingdom) after the microspheres were well dispersed in ultrapure Mili-Q water. Average particle size was expressed as intensity mean diameter.

The amount of IVM entrapped within the microspheres was determined using the following methods. Lyophilized microspheres were washed with 1 ml ethyl acetate three times, vacuum-dried, and then the microspheres and the substances extracted by ethyl acetate were dissolved in 70% ethanol separately and the content of IVM was analyzed using Ultraviolet-Visible spectro-photometer (UNICAM UV 500, UK) at 245 nm.

Heparin Zein Microspheres

The heparin-loaded microspheres were obtained as above described. The vacuum-dried microspheres powder was resuspended in ultrapure Mili-Q water and the heparin-loaded zein microspheres were separated by chromatography to remove unentrapped heparin and characterized with SEM. The encapsulation efficiency and heparin loading were determined using the method described by Smith (15).

Preparation of the Heparin-Loaded Zein Film and IVM-Loaded Zein Tablet

Drug-loaded zein microspheres suspension in 40% ethanol-water solution was dropped onto the surface of the matrix (such as glass or steel) and volatilized at 37°C to form the film.

Drug-loaded zein tablets were prepared by compressing 220 mg of drug-loaded zein microspheres with a mold, and then the tabletted microspheres were placed into wet box at 37°C for 3 days to acquire certain toughness.

The morphologies of the film and the tablet were observed by SEM.

In Vitro Release

In Vitro Release of IVM

IVM-loaded zein microspheres washed with ethyl acetate to remove out unencapsulated IVM and the tablet were placed in PBS buffer (pH 7.4) containing Tween 20 (0.5% w/v), then incubated at 37°C. Medium from each sample was periodically removed and replaced with fresh PBS buffer (pH 7.4) containing Tween 20 or citric acid-NaH$_2$PO$_4$ buffer (pH 2.2) containing pepsin and Tween 20. The IVM content in the release medium and the degradation rate of the tablet were analyzed using an ultraviolet-visible spectrophotometer at 245 and 280 nm, respectively. The percentage of IVM cumulative release (% w/w) and zein degradation rate were investigated as a function of incubation time. Each experiment was performed in six replicates.

In Vitro Release of Heparin

Lyophilized heparin-loaded zein microspheres were washed with ultrapure Mili-Q water to remove out unencapsulated heparin, vacuum-dried and placed into dialysis bag. Then the heparin-loaded zein microspheres and the heparin-loaded zein films were suspended in PBS (pH 7.4), incubated at 37°C, respectively. In vitro half of the medium from each sample was removed and replaced with fresh PBS at a given time interval. The heparin content in the medium was analyzed using the method described by Smith, and the release rate of heparin was calculated.

SEM was used to observe the morphology changes of the microspheres, films and tablets during the release of the drug.

Activity Assays of the Drug

Anticoagulation of Zein Film and Heparin-Loaded Zein Film (16)

This test was divided into four groups: zein film group, Corning culture plate group, physiological saline group and heparin-loaded zein film group. Thrombin, anticoagulant plasma from healthy rabbit and calcium chloride solution were prewarmed at 37°C. Then 0.1 ml plasma, 0.1 ml 0.85% calcium chloride solution and thrombin (100 units/ml) were added into the sample, and began to record the time until cruor. In order to assess the activity of heparin released from zein microsphere film, we also performed this test at the given time during the release of heparin with thrombin time (TT) assay.

Plasma Protein Adsorption and Platelet Adhesion (17)

Fresh healthy human blood was provided by Shanghai Blood Center with approval. The blood was centrifuged at 1000 rpm for 10 min at 4°C to get platelet-rich plasma (PRP) and at 3000 rpm for 10 min at 4°C to get platelet-poor plasma (PPP). The fresh PRP or PPP samples were used in this study. Zein film and heparin-loaded zein microsphere film were sufficiently rinsed with ultrapure Mili-Q water, and immersed into platelet-poor plasma, which was placed at 37°C for 3 hours. After being washed three times, proteins adsorbed on the zein film and heparin-loaded zein microsphere film were removed with 1 wt% aqueous solution of sodium dodecyl sulfate (SDS), and freeze-dried. The protein samples were redissolved in ultrapure Mili-Q water, applied to sodium dodecyl sulfate-polyacrylamide gel electrophoresis (SDS-PAGE).

For evaluation of platelet adhesion, zein film and heparin-loaded film were placed in contact with 50 µl of human platelet-rich plasma (PRP) and left at 37°C for 60-180 min, the number of platelets in the PRP was diluted with PBS to 5×10^6 cells. The PRP was removed and the film was rinsed three times with PBS (18). The adhered platelets were fixed by immersing the film into 2.5% solution of glutaraldehyde in PBS for at least two hours at 4°C. Samples were freeze-dried, then sputter-coated with gold for SEM observation. The number of platelets adhered was determined with colorimetric method (19). Briefly, 150 µl of 4-nitropheny-phosphate disodium salt was added in each well and placed at 37°C for 60 min. The absorption was determined at the wavelength of 405 nm after stopping reaction with 100 µl of 2 N NaOH.

Statistical Analysis

All the data were analyzed by one-way factorial ANOVA and multiple comparisons. Significant effects of treatment were defined using Fisher's PLSD statistic method.

Results and Discussion

HUVECs Compatibility (20)

The biocompatibility was a determinative factor for the application of a new material. In order to develop the use of zein as a coating material, we first evaluated the attachment, spreading and proliferation of HUVECs treated with the zein film and its degraded products. HUVECs attached and adhered onto zein film and the glass plate completely after one hour culture, the percentage of attached cells was more than 90% in each group and no significant differences were observed. Attachment, adhesion and spreading occurred in the first phase of cell/material interactions, and there were no differences between these groups, i.e., both zein film and its degraded product had no effect on attachment and adhesion of HUVECs. Cells cultured in each group maintained their characteristic polygonal morphology, the spreading cells maintained physical contact with each other through filopodia or lamellopodia. However, the morphologies of cells varied greatly on different matrixes, especially on zein film after 3 days of culture (Figure 1). The cytoplasm of HUVECs on zein film was abundant and lacked of thin pinna, which appeared among that of the control group and degraded product group. On the other hand, surface area of cells treated with zein-degraded product was relative smaller than that of the other two groups. And cells did not reach confluency in any group within 5 days.

As shown in Figure 2, at the third day of culture, an enhancement of mitochondrial activity by zein film was observed compared with the control, but not with the zein degraded product group. Significantly higher proliferations were obtained for both zein film and its degraded product group after 5 days of culture. There was no significant difference between zein film group and zein degraded product group.

After being cultured for 2 days with different concentrations of zein degraded product, MTT showed that the growth of HUVECs cultured with zein degraded product was significantly better than the control (n=6, $p<0.05$). A lower amount of addition presented a higher stimulating effect, but the result was not significant (Figure 3). Figure 4 showed the morphology and distribution of HUVECs treated with different concentration of zein degraded product after 2 days of culture by an optical microscope, which had no significant differences.

As shown in Figure 10c, the surface of the zein microspheres film was not smooth because of the existence of microspheres with nanometer to micronmeter. We deduced that at earlier state of the cell proliferation, substratum topography on which cells adhered might be, to some extent, important (21). As a native protein, zein structure contains some hydrone polypeptides. Miyoshi et al. (22) have isolated them from α-zein, identified most of them as tripeptides by Edman degradation and found their inhibiting effect against angiotensin-converting enzyme. In summary, the adhesion and spread of cells weren't affected by zein degraded product, while continually supplemented zein degraded product to the culture medium may play a more important role on the proliferation at later stage, just like some nutrient factors.

Characterization of Microspheres, Film and Tablet

Before processing, zein powder is an unregulated particle. We utilized the phase separation method based on the solubility of zein and drugs to prepare protein microspheres. The microspheres after lyophilization did not aggregated, which can be easily dispersed again in distilled water. The surface of both microspheres encapsulating either IVM or heparin was spherical and smooth with no apparent pores as observed by SEM.

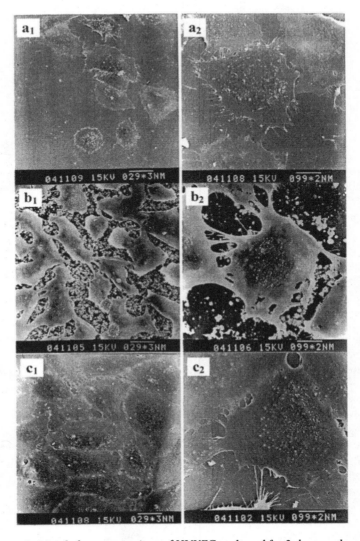

Figure 1. Morphology comparison of HUVECs cultured for 3 days on the glass control (a₁, a₂), zein film (b₁, b₂), and treated with zein degraded product (c₁, c₂, 2 mg/ml) by SEM observation. Bars in left and right columns represent 29 and 9.9 μm, respectively. Reproduced with permission from Reference 20. Copyright 2005 Elsevier B.V.

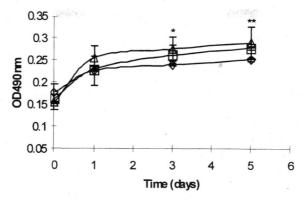

*Figure 2. The proliferation assay of HUVECs on zein film (△), glass control (◇), or treated with zein degraded product (□, 2 mg/ml) after being cultured for 5 days. * Represents a statistical significance between zein film group and the control group; **represents statistical significances compared both zein film group and zein degraded product group with the control group (p<0.05, n=6). Reproduced with permission from Reference 20. Copyright 2005 Elsevier B.V.*

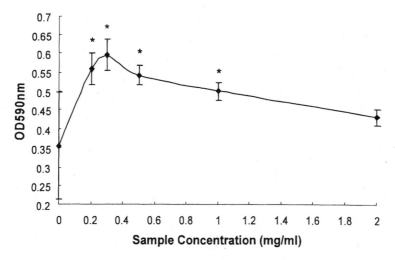

*Figure 3. Effects of zein degraded product on HUVECs proliferation with MTT analysis after being cultured for 2 days (p<0.05, n=6). *Represents significant difference compared with the control (containing 0 mg/ml of zein degraded product). Reproduced with permission from Reference 20. Copyright 2005 Elsevier B.V.*

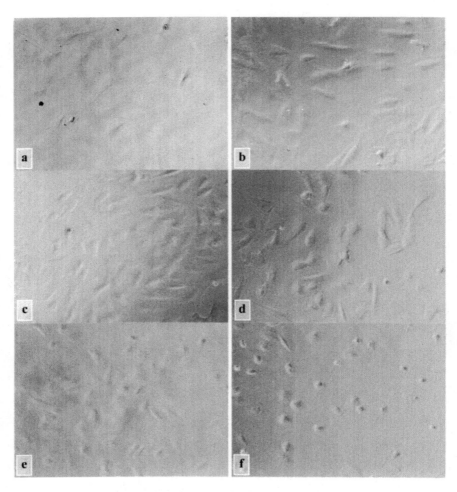

Figure 4. Effect of zein degraded product on HUVECs. Morphology and contribution of HUVECs on Corning microplate after being cultured with different concentrations of degraded product for 2 days were observed by an optical microscope (32× magnification). A: 0 mg/ml (the control group); b: 0.2 mg/ml; c: 0.3 mg/ml; d: 0.5 mg/ml; e: 1 mg/ml; f: 2 mg/ml. (Reproduced with permission from Reference 20. Copyright 2005 Elsevier B.V.)

We observed that the formation of microspheres was affected by many factors (Figure 5 and Figure 6). For example, when concentration of the second solvent, ethanol, was changed, the size or homogeneity of the microspheres changed greatly. When the final concentration of ethanol was under 20%, the protein microspheres could not be formed (Figure 5a); the formed microspheres were heterogeneous when the final concentration of ethanol was from 20% to 40% (Figure 5b); and the optimal concentration of ethanol in final solution was 40%. The homogeneity of the formed microspheres was also affected by adding rate of the second solvent (Figure 6).

In addition, concentration of zein is another important factor affecting the size of microspheres. When the content of zein in 40% of alcohol aqueous solution was 8 mg/ml (23), we got microspheres in diameter of 40-100 nm (Figure 7a); while 500-1500 nm for 20 mg/ml of zein (Figure 7b). One method of the local targeting could be realized by controlling the size of drug delivery systems. When the administration was performed by intravenous injection, the particle size under 1.4 μm could pass through pulmonary circulation; those around 5 μm were accumulated in the lung; while those between 3-12 μm became entrapped within the capillary networks of the lung, liver, and spleen (24).

We also prepared IVM-loaded zein microspheres and heparin-loaded zein microspheres by the same method. The IVM and heparin were as the model of fat-soluble and water-soluble drugs, respectively. As described above, the size of

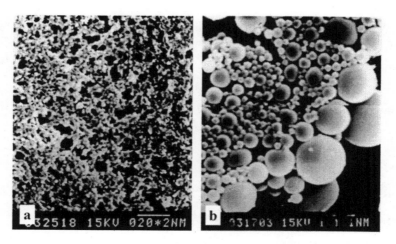

Figure 5. Effect of the second solvent content (ethanol) in the final solution on preparation of microspheres, a: 20%, zein: 20 mg/ml; b: 30%, zein: 20 mg/ml. (Reproduced with permission from reference 26. Copyright 2005.)

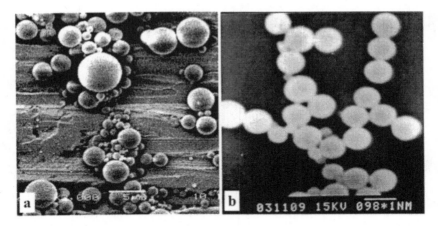

Figure 6. Effect of adding rate of the second solvent (zein: 20 mg/ml) on preparation of microspheres, a: slowly; b: quickly. (Reproduced with permission from reference 26. Copyright 2005.)

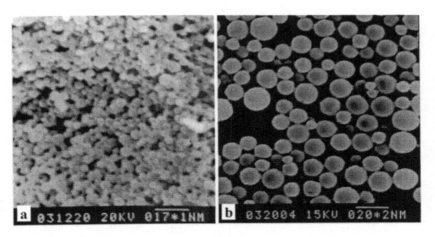

Figure 7. Variety of microspheres sizes prepared by phase separation method, a: zein, 8 mg/ml; b: zein, 20 mg/ml. (a is reproduced with permission from reference 26. Copyright 2005.)

drug-loaded zein microspheres could be controlled by zein concentrations (Figure 8). In addition, the size of zein microspheres increased after loading drugs (Figure 9). The encapsulation efficiencies of IVM and heparin varied with the ratio of zein to both of drugs. Table I showed that the encapsulation efficiency increased with the increase of zein content when the amount of heparin kept constant. However, heparin loading was almost the same, that was to say, total amount of heparin incorporated per zein should be almost the same, so we could control the encapsulation efficiency by zein. On the other hand, the encapsulation efficiency decreased with the increase of heparin content when the amount of zein kept constant. The maximal encapsulation efficiency we have got was 20.4%±2.6% (n=3), limited by the solubility of zein and heparin. As can be seen in Table II, IVM loading and encapsulation efficiency depended on the concentration and the ratio of IVM to zein. Both IVM loading and encapsulation efficiency increased with the increase of IVM and zein concentrations at a given ratio of them; the IVM encapsulation efficiency increased while IVM loading decreased with the decrease of the ratio of IVM to zein. Satisfactory result was obtained when the concentration of IVM and zein was 7.5 and 30 mg/ml, respectively, in which the IVM encapsulation efficiency was nearly 60% and IVM loading was nearly 20%.

The heparin-loaded zein film thickness was 26.3µm (Figure 10a); the film was made from microspheres as observed from the profile (Figure 10b), although some conglutination was observed from the façade (Figure 10c), the size of microspheres was 1-2 µm. One of the most important applications for sustained release material was as the coating material, such as stent coating. Metallic stents exert a continuous radial pressure on the diseased artery, resulting in compression of atherosclerotic plaques, sealing of dissections, and expansion of the vessel. So drug-eluting coating and the initial success of drug-eluting stent captured a great deal of attention and warrants further examination. Now drug-eluting coating aimed at improving the vascular healing response in an attempt to prevent early thrombosis and late neointimal proliferation, is considered as a more logical approach. We could adjust the heparin loading in the film through using heparin-loaded zein microspheres with different encapsulation efficiency, which will affect the release time of the drug to satisfy the need of drug-eluting stents.

The outward and the morphology of the internal structure of the tablet were shown in Figure 10d and 10e. The direct compressing method was applicable for unstable drugs such as peptides because of no heating, but the tabletting process represented the potential risk of damaging microspheres by the mechanical load during compaction (25). In our method, the microspheres after direct compressing and shaping for 3 days were still spherical and smooth without any damage.

222

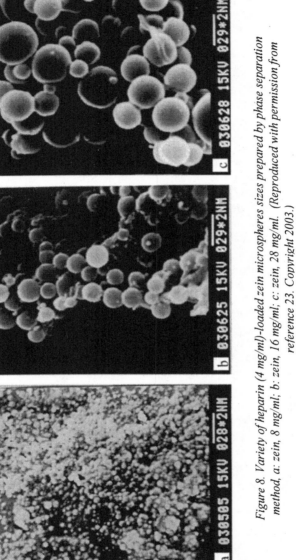

Figure 8. Variety of heparin (4 mg/ml)-loaded zein microspheres sizes prepared by phase separation method, a: zein, 8 mg/ml; b: zein, 16 mg/ml; c: zein, 28 mg/ml. (Reproduced with permission from reference 23. Copyright 2003.)

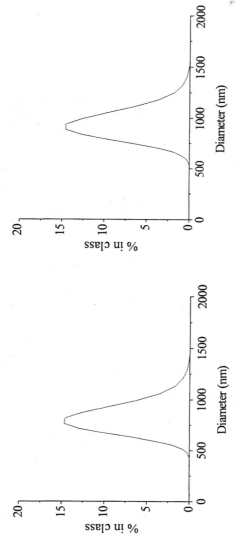

Figure 9. Variety of microsphere sizes before (left) and after (right) IVM loading determined by laser light scattering particle size analyzer.

Table I. Encapsulation Efficiency of Heparin and Heparin Loading in Zein Microspheres (n = 3)

Microspheres Preparation (Concentration of Heparin and Zein)		Encapsulation Efficiency of Heparin(w/w %)[a]	Heparin Loading (w/w %)[b]
Heparin	1.33 mg/ml,	1.97±0.77	0.34±0.17
Zein	4 mg/ml		
Heparin	1.33 mg/ml,	8.74±0.98	0.98±0.11
Zein	8 mg/ml		
Heparin	1.33 mg/ml,	11.94±0.85	0.89±0.06
Zein	12 mg/ml		
Heparin	1.33 mg/ml,	22.77±1.33	1.28±0.07
Zein	16 mg/ml		
Heparin	2.67 mg/ml,	11.94±0.56	1.35±0.06
Zein	8 mg/ml		
Heparin	4 mg/ml,	15.38±1.11	3.46±0.25
Zein	8 mg/ml		
Heparin	5.33 mg/ml,	16.37±0.93	3.67±0.21
Zein	8 mg/ml		
Heparin	4 mg/ml,	16.86±0.74	2.53±0.11
Zein	12 mg/ml		
Heparin	5.33 mg/ml,	19.56±1.30	2.20±0.15
Zein	16 mg/ml		

NOTE: [a] Heparin incorporated into zein microspheres / heparin initially added × 100 %.
[b] Heparin loading (w/w %) = amount of heparin in microspheres/amount of microspheres × 100 %.
SOURCE: Reproduced with permission from Reference 20. Copyright 2005 Elsevier B.V.

**Table II. IVM Encapsulation Efficiency and IVM Loading
in Zein Microspheres (n = 3)**

Preparation Concentrations of IVM and Zein	IVM Encapsulation Efficiency (w/w %)[a]	IVM Loading (w/w %)[b]
IVM 7.5 mg/ml, Zein 30 mg/ml	57.01±0.85	17.16±0.50
IVM 5 mg/ml, Zein 20 mg/ml	50.86±1.28	14.86±0.23
IVM 2.5 mg/ml, Zein 10 mg/ml	47.02±0.43	13.62±0.33
IVM 1.25 mg/ml, Zein 5 mg/ml	38.50±0.31	11.01±0.23
IVM 10 mg/ml, Zein 20 mg/ml	24.73±0.29	15.56±0.18
IVM 7.5 mg/ml, Zein 20 mg/ml	39.86±0.05	16.90±0.56
IVM 2.5 mg/ml, Zein 20 mg/ml	63.02±0.84	9.07±0.24
IVM 1 mg/ml, Zein 20 mg/ml	68.51±0.48	4.05±0.07

NOTE: [a] Encapsulation efficiency (w/w %) = Amount of IVM in microspheres / IVM initially added.
[b] IVM loading (w/w %) = Amount of IVM in microspheres / Amount of microspheres.
SOURCE: Reproduced with permission from Reference 10. Copyright 2004 Elsevier Ltd.

226

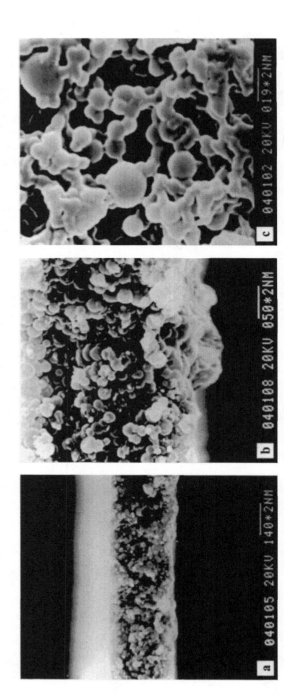

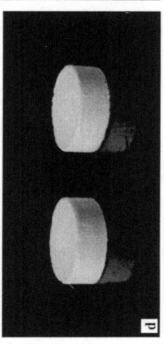

Figure 10. Morphologies characteristics of heparin-loaded zein film (profile: a and b; façade: c. Reproduced from Reference 20. Copyright 2005 Elsevier B.V.) and tablet (outward: d and internal structure: e. Reproduced with permission from reference 10. Copyright 2004 Elsevier Ltd.)

In Vitro Release

In an attempt to develop the applications of zein as the release material, we have examined the in vitro release profiles of both of IVM and heparin in microspheres, films and tablets formulations. The release characteristics were different for IVM and heparin, even both under the same formulation. For the microspheres formulation, both of the releasing curves (Figure 11) indicated the sustained release pattern and appeared biphasic which were characterized by an initial 'burst effect' (phase 1) followed by controlled and sustained release (phase 2). Phase 1 was usually attributed to the drug being present at or near the surface, whereas phase 2 represented the movement of drug entrapped deeper in the zein matrix. The slowing of the release rate probably represented the long diffusion route of drugs. However, for heparin-loaded zein microspheres, there was about 17% of heparin released in 4 hours, about 73% in 48 hours and about only 75% in 72 hours (Figure 11a). At the beginning of the release, because the existence of repellency from heparin on the surface of zein microspheres and continuously massive release, zein microspheres kept spherical; however, with the disappearance of the heparin on the surface after 'burst release', the repellency weakened gradually and microspheres began to aggregate (Figure 12b). For IVM-loaded zein microspheres, there was about 10% of IVM released in 4 hours; only 10% was released every day, and sustained release was kept for 7 days, there was still about 10% of IVM entrapped zein microspheres at last (Figure 11b). Incubation of microspheres in buffer resulted in the similar tendency with heparin-loaded zein microspheres, which aggregated as observed by SEM (Figure 12c) (26). Of course, the micro-channels also began to disappear. Because no digestive enzyme existed, both drugs entrapped into microspheres could not continue to release through diffusion.

The release characteristic of the film appeared also biphasic (Figure 13). In 12 hours, the release of heparin arrived to 33.5%±1.2%, which was characterized by a 'burst effect' (phase 1). Then the release appeared a sustained release (phase 2), after 20 days, the release rates tended to tail off in the subsequent 'slower' phases and only about 54.3%±3.9% of heparin in the film was released at last. Characteristics of heparin-loaded zein film before and after release were also observed by SEM randomly. From the facade of the film, the size of microspheres increased with immersing time of the film in PBS, adjacent microspheres aggregated (Figure 14a-c). From the profile of the film, the microspheres kept spherical, few of them aggregate and the size kept constant (Figure 14d-f) after 1 days, but increased compared with 0 days, which could give explanation to the incomplete release.

Figure 15(A) showed the release of IVM from non-degrading tablet and degrading tablet. Significant difference was observed in release behavior of IVM from tablets with and without pepsin digestion. The tablet without pepsin yielded and IVM release platform during the first 4 days. This might be due to the

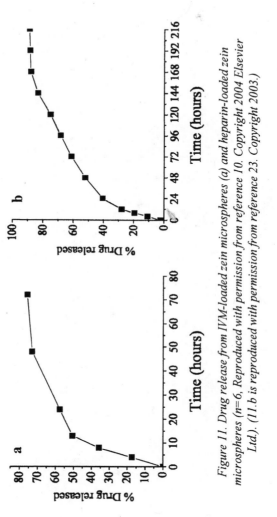

Figure 11. Drug release from IVM-loaded zein microspheres (a) and heparin-loaded zein microspheres (n=6. Reproduced with permission from reference 10. Copyright 2004 Elsevier Ltd.). (11.b is reproduced with permission from reference 23. Copyright 2003.)

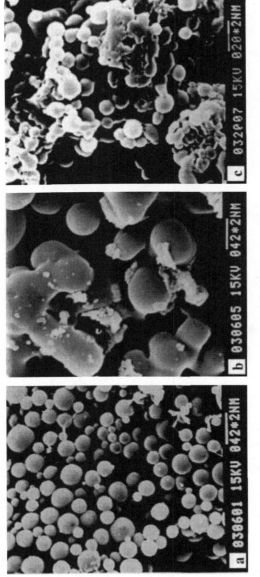

Figure 12. Morphologies of microspheres before (a) and after drug release
(b) heparin loaded zein microspheres for 72 hours; (c) IVM loaded zein microspheres for 24 hours).
(Figures a and b are reproduced with permission from reference 23. Copyright 2003.
Figure c is reproduced with permission from reference 26. Copyright 2005.)

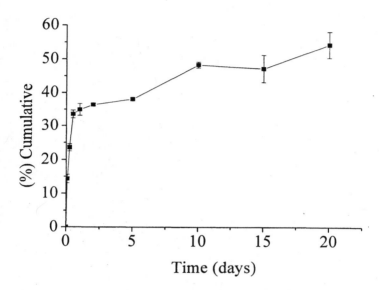

Figure 13. Release kinetics curve of heparin-loaded zein film in PBS (pH7.4, 37°C, n=3). Reproduced with permission from reference 20. Copyright 2005 Elsevier B.V.

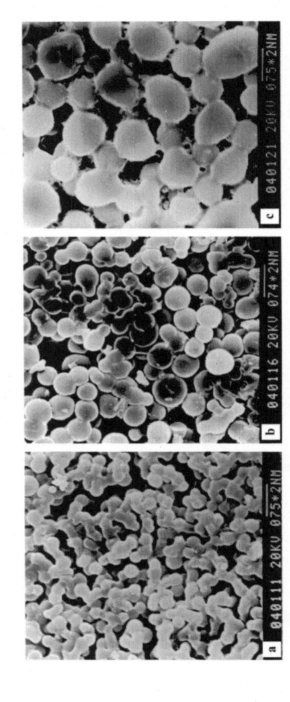

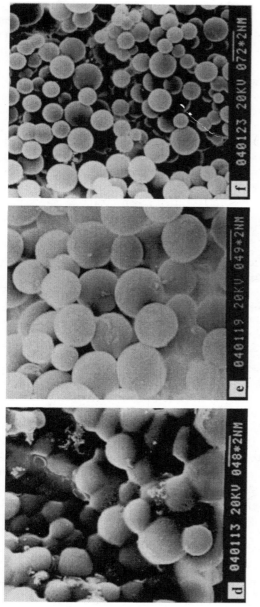

Figure 14. Morphologies of heparin-loaded zein films during release (The façade a: 1 day, bar=7.5 μm; b: 3 days, bar=7.4 μm; c: 6 days, bar= 7.5 μm. The profile d: 1 day, bar=4.8 μm; e: 3 days, bar=4.9 μm; f: 6 days, bar=7.2 μm). Reproduced with permission from reference 20. Copyright 2005 Elsevier B.V.

234

hydrophobic nature of zein and IVM, and the internal structure with a thick and dense wall. Normally, it needed longer time for water to penetrate through more hydrophobic zein matrix and across the thick wall to let IVM diffuse out. For the enzymatic degraded tablet, the release of IVM was almost linear with the time (zero-order release), and continued for 11 days. It was generally recognized that the drug release from matrix tablet depends on diffusion that was influenced by degradation of the matrix (27). The release of IVM from the tablet could maintain a constant drug concentration in plasma and a constant pharmacological effect in vivo. Figure 15(B) showed pepsin-degraded rate of tablets. After 8 days incubation, about 50% of the tablet were degraded, which reached to 100% after 13 days. So it was possible that the tablet could endure the gastric degradation after oral administration. This drug delivery system might also be used to administer drugs topically in most excessive environment, such as ruminal bolus or subcutaneous implant. The drug release from it depended on degradation and diffusion, which could achieve zero-order release, which eliminated the need for surgery to remove the device after drug releasing (28).

In summary, because of the aggregation of zein in water solution, the channels or pores to the surfaces of microspheres, films and tablets disappeared during the release of the drugs, which caused that some drugs entrapped deeper in the structure could not be released without the action of the hydrolases. So the releases of both drugs from microspheres, films and tablets in vitro all characterized with no erosion-incomplete release.

Activity of the Drug Released

There was an important problem remained to be answered, i.e., did the drug after processing still keep its high activity? In this study, we examined the hemocompatibility of heparin-loaded zein films by TT test, protein adsorption and platelet adhesion assay. As Table III showed, there were no significant differences between the control group, physiological saline group and zein film group for the anticoagulation, the TT time was controlled within 30 s approximately. On the other hand, heparin-loaded zein microsphere film group showed a significant anticoagulation and the plasma kept incoagulable all the time. This result indicated that zein film itself didn't prevent acute or subacute thrombosis if coated directly on cardiovascular devices such as stent without seeding endothelial cells before implantation. However, for heparin-loaded zein film, which could be obtained at low temperature, heparin kept the high activity.

We also designed a simplified test to determine anticoagulating effect of heparin-loaded zein film during the release of heparin (Table IV). As cruor always occurs in several hours after implantation of stent, it implies that heparin-loaded zein microsphere film could maintain the anticoagulant activity by slow heparin release and zein degradation in vivo.

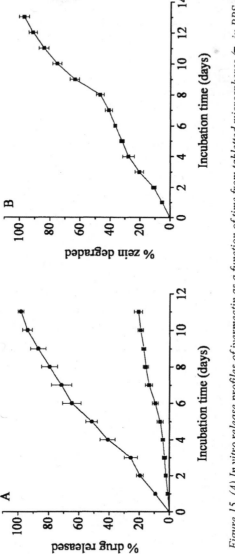

Figure 15. (A) In vitro release profiles of ivermectin as a function of time from tabletted microspheres (■, in PBS, pH 7.4, 37°C) and from pepsin degraded tabletted microspheres (●, in citric acid-NaH₂PO₄ buffer, pH 2.2, 37°C). (B) In vitro degradation of tabletted microspheres with pepsin in citric acid-NaH₂PO₄ buffer, pH 2.2, 37°C. Each point represents the mean ± SD (n = 6). Reproduced with permission from reference 10. Copyright 2004 Elsevier Ltd.

236

Table III. Anticoagulation of Zein Films and Heparin-Loaded Microspheres Films (n=3)

Groups	TT(s), Average±SD
Control	30.33±5.13
Physiological saline	29.33±5.51
Zein film	29.33±3.21
Heparin-loaded zein film	Incoagulable*

NOTE: "*" represents significant difference compared with the control ($p<0.01$).
SOURCE: Reproduced with permission from reference 20. Copyright 2005 Elsevier B.V.

Table IV. The Activity of Heparin-Loaded Zein Microspheres Film During Release of Heparin (n=3)

Time of Heparin Release	Time of Formation of Fibrinclot	Time of Clotting of Plasma	Significant Difference
Control	-	20 ± 2.5 s	-
2 h	-	Incoagulable	**
6 h	-	Incoagulable	**
12 h	7.3 ± 1.2 min	12 ± 1 h	**
24 h	2.7 ± 0.7 min	36.7 ± 11.5 min	**
48 h	-	42 ± 11 s	*
96 h	-	48 ± 4.6 s	*
240 h	-	54 ± 13.3 s	*

NOTE: "*" represents statistical significance compared with the control at $p<0.05$, and "**" at $p<0.005$.
SOURCE: Reproduced with permission from reference 20. Copyright 2005 Elsevier B.V.

Platelet adhesion on films from human plasma is an important test for the evaluation of the blood compatibility of materials. Figure 16 showed that the relative number of platelets adhered on both the zein film and heparin-loaded zein film decreased significantly compared with that on glass plate ($p<0.05$), only about 12% and 22% of it, respectively. Moreover, numbers of adhering platelets on heparin-loaded film was the minimum and significantly lower than zein film. While the morphology of platelets varied on these surfaces. Platelet adhered, spreaded and developed characteristic pseudopodia on the glass control (Figure 17a), but little platelet adhesion was observed on the zein film (Figure 17b), especially heparin-loaded zein film (Figure 17c). The platelets adhered to zein films maintained their original round shape compared with those adhered to the glass control.

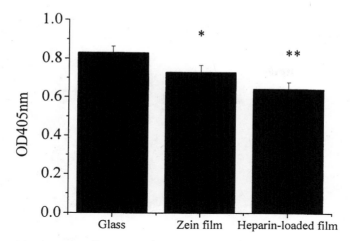

*Figure 16. Platelet adhesion on different material surfaces. Data were expressed as means ±standard deviations. * Represents a statistical significance compared with the control group; ** represents statistical significances compared with both the control group and zein film group (p<0.05, n=6). Reproduced with permission from reference 20. Copyright 2005 Elsevier B.V.*

The platelet adhesion mainly turns on the types and conformation of adsorbed plasma protein; plasma protein adsorption strongly depends on their surface characteristics, e.g. surface hydrophobicity or charge. The higher the particle surface charge density, the more proteins are adsorbed which may be mainly driven by Coulomb forces (29,30). Hydrophobic surface has been indicated to be more suitable to protein adsorption than hydrophilic surface (31). Hydrophobic property of zein might contribute to its high adsorbing ability to proteins from plasma, compared with the glass control. Human serum albumin was the predominant adsorbed protein, as indicated by SDS-PAGE (data not shown). The higher content of albumin adsorbed on the zein film might result in reduced platelet adhesion (32). While much lower platelet adhesion of heparin-loaded zein film might be due to the electrostatic repulsion between negative charges of platelet and heparin released from the film (33). All the possibilities caused that platelet adhered on the heparin-loaded zein film easily desquamated; so fewer cells could be observed in Figure 17c, but the OD value still got to 0.6 as shown in Figure 16. We considered that some platelets might strip away during sample preparation for SEM analysis.

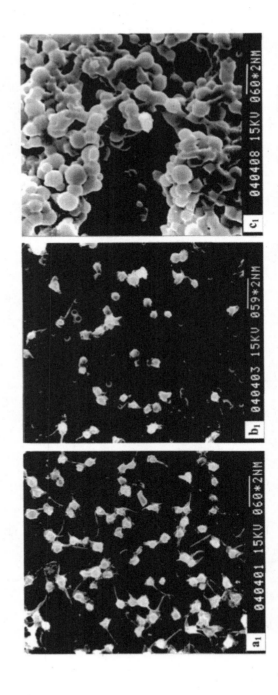

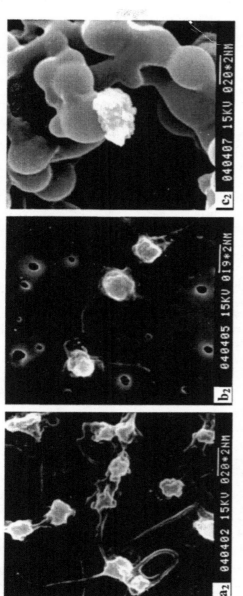

Figure 17. SEM observations of blood platelets adhered on different matrixes for 60 min: the glass control (a_1, a_2); zein film (b_1, b_2) and heparin-loaded zein film (c_1, c_2). Bars in left and right column represent 6 and 2 μm, respectively. Reproduced with permission from reference 20. Copyright 2005 Elsevier B.V.

Conclusion

Zein, a native protein, and its degraded product showed a good biocompatibility on the growth of HUVECs, which gave us the promise of improving biocompatibility of cardiovascular devices after modification with endothelial cells. In addition, we could prepare drug-loaded zein microspheres by phase separation method and the size could be controlled, which might realize targeting of the drugs to some extent. The encapsulation efficiency could be controlled based on the variety of the ratio of zein to the drug, but that of the water-soluble drug (22.77±1.33) was lower than that of the fat-soluble drug (68.51±0.48). The film and tablet from drug-loaded zein microspheres were prepared under non-destructive conditions, the drugs still kept high activity. The drugs released from film and tablet could be slowed for more than 20 days (remaining 50%) and 11 days (remaining 80%), respectively, which might depend on the diffusion of drugs through the matrix. It is notable that IVM tablet achieved zero-order release with pepsin action.

Acknowledgement

This study was supported by the National Program on Key Basic Research Projects of China (973 Program, 2005CB724306), the National Hi-Tech Research and Development Plan of China (863 Project, 2002AA327100), the National Natural Science Foundation of China (30270365, 30470477) and the Chinese Academy of Sciences (BaiRenJiHua).

References

1. Hatefi, A.; Amsden, B. *J. Controlled Release* **2002**, *80*, 9-28.
2. Vasir, J.K.; Tabwekar, K.; Garg, S. *Int. J. Pharm.* **2003**, *255*, 13-32.
3. Giunchedi, P.; Juliano, C.; Gavini, E.; Cossu, M.; Sorrenti, M. *Eur. J. Pharm. Biopharm.* **2002**, *53*, 233-239.
4. Mao, S.R.; Chen, J.M.; Wei, Z.P.; Liu, H.; Bi, D.Z. *Int. J. Pharm.* **2004**, *272*, 37-43.
5. Lim, S.T.; Martin, G.P.; Berry, D.J.; Brown, M.B. *J. Controlled Release* **2002**, *66*, 281-292.
6. Sahin, S.; Selek, H.; Ponchel, G.; Ercan, M.T.; Sargon, M.; Hincal, A.A.; Kas, H.S. *J. Controlled Release* **2002**, *82*, 345-358.
7. Latha, M.S.; Lal, A.V.; Kumary, T.V.; Sreekumar, R.; Jayakrishnan, A. *Contraception* **2000**, *61*, 329-334.
8. Lee, A.C.; Yu, V.M.; Lowe III, J.B.; Brenner, M.J.; Hunter, D.A.; Mackinnon, S.E.; Sakiyama-Elbert, S.E. *Exp. Neurol.* **2003**, *184*, 295-303.
9. Mori, H.; Tsukada, M. *Rev. Mol. Biotechnol.* **2000**, *74*, 95-103.

10. Liu, X.M.; Sun, Q.S.; Wang, H.J.; Zhang, L.; Wang, J.Y. *Biomaterials* **2005**, *26*, 109-115.

11. Dong, J.; Sun, Q.S.; Wang, J.Y. *Biomaterials* **2004**, *25*, 4691-4697 & *Materials Today* 2004, July/August, 24.

12. Gong, S.J.; Wang, H.J.; Sun, Q.S.; Xue, S.T.; Wang, J.Y. *Biomaterials* **2006**, *27*, 3793-3799.

13. Sun, Q.S.; Dong, J.; Lin, Z.X.; Yang, B.; Wang, J.Y. *Biopolymer* **2005**, *78*, 268-274.

14. Jaffe, E.A.; Nachman, R.L.; Becker, C.G.; *J. Clin. Invest.* **1973**, *52*, 2745-2756.

15. Smith, P.K.; Mallia, A.K.; Hermanson, G.T. *Anal. Biochem.* **1980**, *109*, 466-473.

16. Lin, W.C.; Liu, T.Y.; Yang, M.C. *Biomaterials* **2004**, *25*, 1947-1957.

17. Keuren, J.F.W.; Wielders, S.J.H.; Willems, G.M.; Morra, M.; Cahaland, L.; Cahalan, P.; Lindhout, T. *Biomaterials* **2003**, *24*, 1917-1924.

18. Xu, J.P.; Ji, J.; Chen, W.D.; Fan, D.Z.; Sun, Y.F.; Shen, J.C. *Eur. Polym. J.* **2004**, *20*, 291-298.

19. Bellavite, P.; Andrioli, G.; Guzzo, P.; Arigliano, P.; Chirumbolo, S.; Manzato, F.; Santonastaso, C. *Anal. Biochem.* **1994**, *216*, 444-450.

20. Wang H.J.; Lin Z.X.; Liu X.M.; Sheng S.Y.; Wang J.Y. *J. Controlled Release.* **2005**, 105, 120-131.

21. Rajnicek, A.M.; McCaig, C.D. *J. Cell Sci.* **1997**, *110*, 2915-2924.

22. Miyoshi, S.; Ishikawa, H.; Kaneko, T.; Fukui, F.; Tanaka, H.; Maruyama, S. *Agric. Biol. Chem.* **1991**, *55*, 1313-1318.

23. Wang H.J.; Wang L.S.; Wang J.Y. *Sci. Tech. Engng.* (in Chinese) **2003**, *3*, 557-560.

24. Thakkar, H.; Sharma, R.K.; Mishra, A.K.; Chuttani, K.; Murthy, R.R. *AAPS Pharm. Sc. iTech.* **2005**, *6*, E65-E73.

25. Kühl, P.; Mielck, J.B. *Int. J. Pharm.* **2002**, *248*, 101-114.

26. Sun Q.S.; Lin Z.X.; Liu X.M.; Wang J.Y. *Heilongjiang Anim. Sci. Vet. Med.* (in Chinese) **2005**, *10*, 13-15.

27. Zuleger, S.; Lippold, B.C. *Int. J. Pharm.* **2001**, *217*, 139-152.

28. Zimmerman, G.L.; Mulrooney, D.M.; Wallace, D.H. *Am. J. Vet. Res.* **1991**, *52*, 62-63.

29. Tanaka, M.; Motomura, T.; Kawada, M.; Anzai, T.; Kasori, Y.; Shiroya, T.; Shimura, K.; Onishi, M.; Mochizuki, A. *Biomaterials* **2000**, *21*, 1471-1481.

30. Gessner, A.; Lieske, A.; Paulke, B.R.; Muller, R.H. *Eur. J. Pharm. Biopharm.* **2002**, *54*, 165-170.

31. Tangpasuthadol, V.; Ponchaisirikul, N.; Hoven, V.P. *Carbohydr. Res.* **2003**, *338*, 937-942.

32. Abraham, G.A.; de Queiroz, A.A.A.; Roman, J.S. *Biomaterials* **2002**, *23*, 1625-1638.

33. Kang, I.K.; Kwon, O.H.; Kim, M.K.; Lee, Y.M.; Sung, Y.K. *Biomaterials* **1997**, *18*, 1099-1107.

Chapter 13

Antibody Modified Collagen Matrix for Site-Specific Gene Delivery

Xu Jin[1], Lanxia Liu[1], Lin Mei[1], Xigang Leng[1], Chao Zhang, Hongfan Sun[1], Cunxian Song[1,*], and R.J. Levy[2,*]

[1]Institute of Biomedical Engineering, Chinese Academy of Medical Sciences and Peking Union Medical College, The Tianjin Key Laboratory of Biomaterial Research, Tianjin 300192, People's Republic of China
[2]The Children's Hospital of Philadelphia, Abramson Pediatric Research Center, Suite 702, 3516 Civic Center Boulevard, Philadelphia, PA 19104

We investigated the feasibility of immobilizing plasmid DNA on collagen matrix through covalently coupled anti-DNA antibody to achieve long-lasting and site-specific DNA delivery. Methods: Collagen was modified with SPDP (N-succinimidyl-3- (2-pyridyldithiol)-propionate) / DTT to introduce –SH group on collagen matrix. A pyridyl disulfide modified anti-DNA IgM wasthen covalently linked with collagen through thiol reaction. These antibodies enabled tethering of plasmid DNA through highly specific antigen-antibody affinity. Radioactive-labelled antibody and pDNA were used to evaluate binding capacity and stability. A plasmid EGFP-N3 was tethered on anti-DNA antibody-modified collagen-coated coronary stent as model gene delivery system that was assessed in both cell culture and in a rabbit model. Results showed that the amount of antibody bound on collagen films was 15 times higher than that of the control films (without SPDP chemical linking), and that the retention time of the antibody on films was also significantly longer. The pDNA-tethered and collagen-coated stents had no detrimental effect on cell growth. In cell culture studies, numerous GFP-transduced cells were detected only on the stent surface, thus demonstrating efficient and highly localized gene delivery. The overall GFP transduction efficiency in treated rabbit coronary arteries was 2.8±0.7% of the total cells. However, the rate of neointima transduction was about 7% of total cells in this region. Importantly, no distal spread of the vector was detectable by PCR, either in distal organs, or in the downstream segments of the stented arteries. In conclusion, we report the first successful use of anti-DNA antibody-modified stent as plasmid gene delivery system; this technique thus represents an efficient and highly localized gene delivery system both *in vitro* and *in vivo*.

The success of gene therapy for cardiovascular diseases, such as restenosis, requires an appropriate gene vector and a device to deliver the vector into the diseased vascular site *(1)*. Vector targeting to a specific tissue or cell type would enhance gene therapy efficacy and permit the delivery of lower doses, which should result in reduced toxicity *(2)*. Vector localization has specific advantages for gene therapy of vascular obstruction and restenosis *(1-4)*. Successful gene transfer from gene-delivery balloon expandable stainless-steel stents has been demonstrated in a number of animal model studies *(5-9)*. However, in most of these studies the gene vectors (viral or nonviral) were physically adsorbed to the stent surface by dipping or spraying the stent. Vectors attached by such means are easily eluted and spread via systemic circulation to distal organs, which resulted in low transduction efficiency and high side effects. For this purpose, our group's previous studies have focused on covalently linking an anti-adenoviral antibody to collagen coatings on the stents using a bifunctional crosslinker SPDP, followed by adenoviral vector tethering through highly specific immuno-conjugation between the immobilized anti-adenoviral antibody and the incoming adenoviral vector *(5)*. This approach was demonstrated to be feasible, with high levels of regional arterial expression and no detectable spread of vector beyond the implanted artery *(5, 10-12)*.

An ideal vector would offer both efficient transduction and long-term transgene expression in target cells, yet without the risk of systemic distribution, immunogenicity, or cytotoxicity *(4)*. Unfortunately such an ideal vector does not yet exist. Plasmid DNA (pDNA) is a popular nonviral vector that has been successfully used in experimental gene therapy and clinical trials *(13-17)*. However, pDNA results in much less extensive gene transfer than viral vectors do. Since pDNA has major safety advantages over viral vectors, methodology to improve the efficiency of pDNA transduction is of great importance. There have been numerous attempts to improve pDNA transduction *(18-21)*, with the most promising avenue of research is cationic liposomes *(22)*. While many cationic liposomes can facilitate pDNA transduction into various cells *in vitro*, only a few cationic lipid/pDNA complexes have shown efficient gene transduction without causing cytotoxicity *(23-26)*. Moreover, cationic liposomes could not facilitate the nuclear entry of pDNA, which is the most important step for pDNA expression in target cells. Previous investigation by our group has demonstrated that a novel pDNA vector composed of plasmid DNA / anti-DNA Antibody /Cationic lipid could enhance pDNA transduction efficacy both *in vitro* and *in vivo(27)*. The new vector system allowed use much lower dosage of cationic lipid in the triple complex than that of in the traditional cationic lipid/pDNA complexes. We thought to combine the advantages of this novel nonviral vector with our novel delivery system of antibody-modified stent to achieve highly efficient intravascular site-specific pDNA transduction. Thus the objectives of the present study were to formulate a novel stent-based pDNA delivery system using the collagen-coating reaction and to evaluate its feasibility and efficacy *in vitro* and *in vivo*. We intended to further investigate the binding stability and the controlled delivery behavior of vectors bound on collagen coatings through the anti-body immobilization approach. Also this study was focused on using collagen matrix as a general platform, examining their localized gene delivery characteristics as a common interest.

Results

Binding Capacity, Stability and Release Kinetics of ^{125}I-labeled Anti-DNA Antibody on the Collagen Films

The anti-DNA antibody was successfully labeled with [125]I. The unbound free radioisotope ^{125}I was completely separated from ^{125}I-labled antibodies by gel filtration on a Sephadex G-50 column, as indicated in Figure 1A. The radiochemical purity reached a level of 98%, as determined by paper chromatography. The concentration of ^{125}I-Antibody-SPDP was about 60mg/µl, as determined by the Coomassie Brilliant Blue G-250 dye-binding method. The amount of antibodies bound on the SPDP activated collagen film (made from 1.0mg dry collagen) was 15 times higher than on the control collagen films without the SPDP chemical linking (Figure 1B, $p<0.01$). The chemically bound ^{125}I-labeled anti-DNA antibody on collagen films showed a sustained release over the course of 16 days and 25% of the total bound antibodies were retained on the films by the end of 16 days. However, the physically absorbed antibody on the control films was almost completely eluted out within 4 days (Figure 1C).

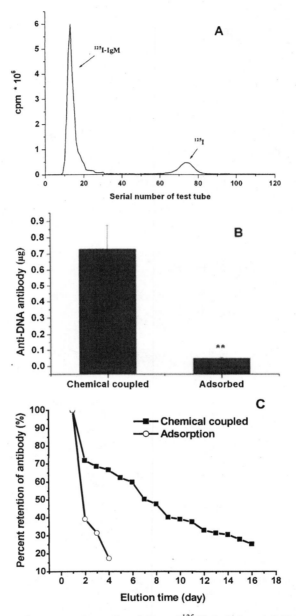

Figure 1. Binding capacity and stability of ^{125}I labeled anti-DNA antibody on collagen films:(A) Purification of ^{125}I-labeled antibody; (B) The binding capacity of antibody on collagen films; (C) The binding stability of antibody on collagen films

Binding Capacity, Stability and Release Kinetics of pDNA Tethered on Anti-DNA Antibody Immobilized Collagen Films

The amount of pDNA tethered on the anti-DNA antibody immobilized collagen films was about 2-fold higher than on the control films with physically-absorbed anti-DNA antibody (p<0.05), as indicated in Figure 2A. The in vitro pDNA elution study (binding stability experiments) showed that about 30% of pDNA released within the first day followed by a sustained release lasted for more than 13 days. However, the pDNA was released from control films much more rapidly, and no detectable radioactivity was seen after 5 days (Figure 2B).

Collagen-Coated Coronary Stent Mediated Site-specific Plasmid DNA (pDNA) Delivery in Vitro

The plasmid DNA was tethered on collagen-coated coronary stents in the same way that of on the collagen films. In cell culture, the stents showed no detrimental effect on cell growth. The antibody immobilized and *pEGFP-N3* tethered stent demonstrated efficient gene transduction in the rat arterial smooth muscle cells (A10 cells). The stents retrieved from cell culture after 72 hours of incubation showed numerous GFP-transduced cells that exclusively infiltrated the collagen coatings on stent, indicating a highly localized and efficient gene delivery pattern (Figure 3A). However, the control stents, immobilized with nonspecific antibody and incubated with same amount of *pEGFP-N3*, resulted in very few transduced GFP positive cells on the stent (Figure 3B).

In Vivo Gene Delivery

The antibody-immobilized/pEGFP-N3 tethered stents were retrieved from rabbits after 7 days of implantation. These stents showed significant cell deposition and widespread, active GFP transduction as shown in Figures 4A and 4B. The artery segment that had been in close contact with the antibody-immobilized and pEGFP-N3 tethered stent showed strong fluorescent signal in the neointimal and medial layers. In contrast, in the artery segments treated with control stents, only weak autofluorescence can be detected in the neointima and media (Figure 4C). Morphology studies demonstrated an overall transduction efficiency of 2.8±0.7% of total cells in modified stent. However, neointimal transduction levels were about 7% of total neointimal cells and medial transduction levels were more than 3% of medial cells. Adventitial cells were hardly transduced (Figure 4D). Immunohistochemical findings using an anti-GFP antibody correlated well with the findings from fluorescent microscopy (Figure 4E). In contrast, the control vessels were negative for anti–GFP antibody staining (Figure 4F). PCR analyses of homogenates from distal organs (lung,

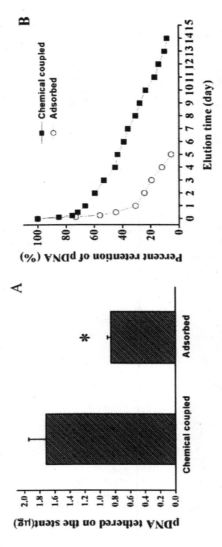

Figure 2. Loading capacity and stability of the chemo-immunity bound plasmid DNA on anti-DNA antibody modified collagen films compared with control films with physically absorbed pDNA. (A) The total amount of pDNA bound on collagen films ($p<0.05$). (B) Results from in vitro elution study.

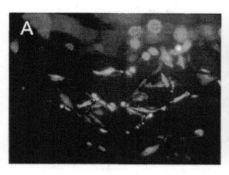

Figure 3. pEGFP transduction in cell culture (rat arterial smooth muscle cell) induced by collagen-coated coronary stents (A) Stent with pEGFP tethered through chemically linked anti-DNA antibody (FITC; 111.8×). (B) Negative control stent without antibody modification (FITC; 111.8×).

liver, kidney) and downstream arterial wall samples, obtained at the time of termination of the experiment, confirmed the absence of GFP DNA in these tissues (Figure 4G). Together, these data strongly indicate that this novel gene delivery system achieved highly efficient and site-specific gene transduction.

Discussion

Recently, more and more attention has focused on localized gene therapy for vascular diseases such as arteriosclerosis, arterial restenosis, vein graft disease, systemic and pulmonary hypertension and cerebrovascular disease *(4, 28, 29)*. All of these strategies require the combination of a therapeutic gene product, an appropriate vector as well as a device for delivering the vectors to the target cells. Most of these therapeutic methods are based on viral vectors and catheter vascular gene delivery strategies. Viral vectors, including adenovirus, retrovirus and adeno-associated virus are the most common for high transduction efficiency in dividing and nondividing cells. However, a number of disadvantages, such as immune response and limited DNA insert size, have impeded the application of these virus-based systems. Therefore, more attention has been focused on nonviral vectors including pDNA and pDNA/liposome to obtain reduced immunogenicity and high level of safety.

Previous studies by Levy's group *(27)* have reported a novel nonviral vector, composed of plasmid <u>D</u>NA /anti-DNA <u>A</u>ntibody /<u>C</u>ationic lipid triplexes (pDAC), which exhibits high transduction efficiency, high nuclear entry and low toxicity. The pDAC vectors formed stable nanoscale micelles with a mean particle diameter of around 360nm through a process of self-assembly. Once

250

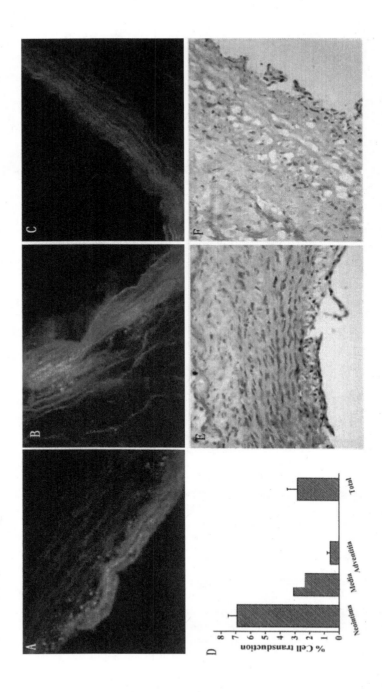

```
G 1 2 3 4 5 6 7 8
```

Lane 1 DNA ladder;
Lane 2 positive control;
Lane 3 lung;
Lane 4 liver;
Lane 5 kidney;
Lane 6 proximal 3cm CCA;
Lane 7 femoral artery;
Lane 8 negative control.

Figure 4. Rabbit carotid arterial samples retrieved after 7 days stenting. (A&B) Examples of GFP expression following implantation of the stent with pEGFP tethered through chemically linked anti-DNA antibody. (FITC; 146.7×). (C) Example of control arteries following implanting control stent (FITC; 146.7×). (D) Results from morphometry study (p<0.05). (E) Immunohistochemical confirmation of GFP expression in the arteries stented with the pEGFP tethered and anti-DNA antibody modified collagen-coated stents (immunoperoxidase; 293.4×).(F) Negative immunoperoxidase activity in control stented arteries (using non-specific IgM, 293.4×). (G) Vector biodistribution assessed with PCR

formulated, the pDAC nonviral vector could achieve stable physical characteristics and high transduction efficiency. The studies of Avramesa et al. showed that anti-DNA antibodies could penetrate into various types of living cells, enter the nucleus; yet they did not appreciably alter cell viability or metabolism *(30)* Anti-DNA antibodies used for site-specific gene delivery applications are of high purity and low dose (<1μg); also, the amount of unbound antibody can be minimized by immobilization. Thus, the risk for anti-DNA antibody-related side effects should be hypothetically low.

Although some new methods of catheter infusion have been developed *(1)*, such as double-balloon catheter, dispatch catheter and the "needle injection" catheter, the weakness inherent in catheter-based vascular gene delivery systems, including the lack of sustained release, localization and the propensity for systemic vector distribution, may be largely overcome using a stent-based gene delivery system. Stents provide a versatile platform for intravascular site-specific gene delivery because of their long-term residence within the vessel. In 1998, Yu et al. first reported significant expression of the LacZ reporter gene in vascular tissue *(8)*. Since then, successful gene transfer from gene-delivery stents has been demonstrated in a number of animal model studies. However, most of these studies reported the bioactive gene vectors were attached to the stent surface by dipping or spraying the stent. In theory, washed out by artery blood flow, vectors attached by dipping or spraying may easily spread along systemic circulation to distal organs, thereby resulting in significantly reduced gene transduction efficiency. Perhaps this is the reason why such a high dose of vector was administrated in many studies. In the mean time, side effects will appear with the spread of vectors. Previous studies by our group reported a novel stent-based gene delivery system utilizing collagen-coatings on the stent *(5)*. The collagen provided a reactable platform for the afterwards chemical immobilization of anti-viral antibody and adenovirus tethering; this approach achieved highly-efficient and site-specific gene transduction without any distal spread. Several research groups have reported that pDNA delivery from a poly (lactic-co-glycolic acid) (PLGA)-coated stent achieved gene transduction in vivo *(6, 7)*. However, these studies showed that part of polymer-DNA fragments spread along systemic circulation to distal tissues or organs.

In the present study, we investigated the hypothesis that pDNA could be coupled on the collagen coatings of coronary stents by the combination of chemical immobilization and immuno conjugation approach. An anti-DNA antibody was immobilized on the collagen coatings through chemical linkage. The pDNA was tethered onto the collagen-stents surface via immuno affinity between the anti-DNA antibody and the pDNA molecules. Our preliminary experiments showed that numerous transduced cells only infiltrate the collagen coatings on stent both in cell culture and in vivo and exhibit no distal spread of vector, indicating a highly localized and efficient gene delivery.

A hetero-bifunctional crosslinker, N-Succinimidyl-3-(2-pyridyldithiol)-propionate (SPDP), was used to chemically link the anti-DNA antibody with the metal stent through collagen coatings that functioned as a source of reactive amine groups. The basic chemical reactions for antibody binding on the collagen coatings include: 1) to introduce –SH function group on the collagen; 2) to introduce pyridyl disulfide groups on the antibody; and 3) to couple the thiolated antibody to the –SH activated collagen coatings. Thus, the anti-DNA antibody was chemically immobilized onto the collagen-coated stent. These reactions were illustrated in Figure 5. The ^{32}P- labeled pDNA was used to precisely quantify the amount of DNA linked on collagen-coated stent and to qualitatively characterize the release kinetics of pDNA from the stent. By using the ^{32}P-labeled pDNA, we were able to directly assay the loading capacity of pDNA on stent with high sensitivity. We selected an anti-DNA monoclonal IgM antibody that recognizes both double- and single-stranded DNA, thus ensuring a wide variety of therapeutic genes that could be tethered on stents, and independent of the pDNA size.

During the past several years, our understanding of the pathogenesis of restenosis has advanced rapidly *(31)*. Although smooth muscle cell proliferation is probably the main mechanism of restenosis after coronary stenting, factors such as thrombosis, platelet activation, vasoconstriction, leukocyte adhesion, matrix deposition and remodeling are also important pathogenetic mechanisms in other settings. More than 20 kinds of therapeutic genes can be used to achieve pleiotropic effects, such as TIMP *(32)* for inhibiting SMC migration, TFPI for inhibiting thrombosis, VEGF *(13)* and iNOS *(33)* for accelerating endothelialization and some anti-tumor genes for inhibiting proliferation of SMC. Antisense DNA *(34-38)* or siRNA can also be used for this purpose. Jeschke et al. revealed that a combination of multiple genes was feasible and might be more effective than a mono-therapy *(39)*. Based on the characteristics of the anti-DNA antibody, we hypothesized that this anti-DNA antibody modified stent might be able to tether multiple genes on a single stent. This "cocktail approach" may be necessary to affect multiple mechanisms, and therefore to achieve better therapeutic effects.

Although the results of the present study are very promising, we further hypothesize that metal stents could covalently link functionalized amine group through a chelation reaction, thus without collagen coating; such an approach would avoid the potential side effects of collagen surface and might produce better results.

Conclusion

We report for the first time the successful use of anti-DNA antibody immobilized metal stent as gene delivery system; we demonstrate efficient and highly localized gene delivery to arterial smooth muscle cells. We thus conclude that a vascular stent is indeed a suitable platform for intravascular site-specific pDNA delivery without systemic spread of vector. Furthermore, the same design

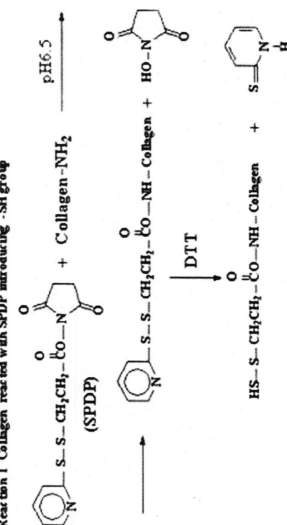

Reaction 1 Collagen reacted with SPDP introducing -SH group

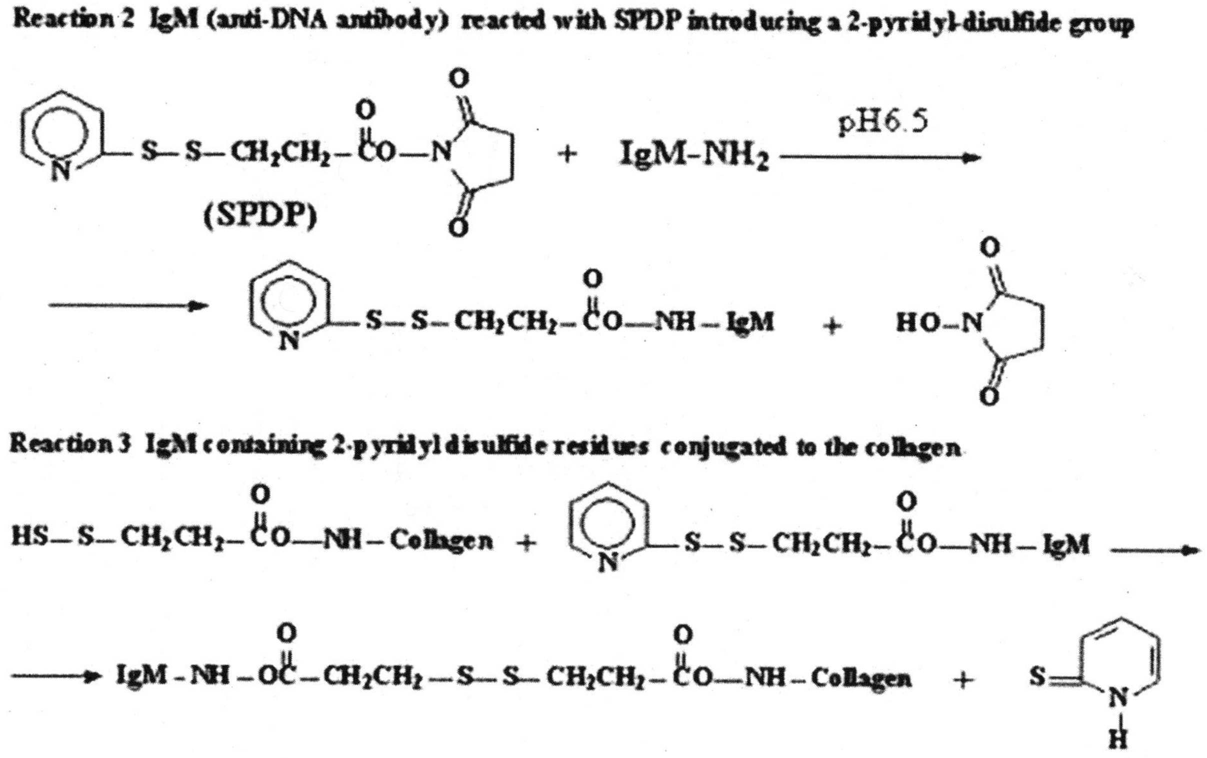

255

Figure 5. Scheme of the reactions involved in anti-DNA antibody linking on collagen films through SPDP activation.

could be used on the surface of various medical devices such as leads, coils, vascular grafts and heart valves, to deliver therapeutic polynucleotides to targeting sites for achieving efficient gene therapy.

Materials and Methods

Materials

Bovine dermal type I collagen (vitrogen 100®, 3.0mg/ml in 0.012N HCl solution) was obtained from Vitrogen Cohesion Technologies (Palo Alto, CA, USA). The EGFP-N3 plasmid was purchased from Clontech (Palo Alto, CA, USA). A monoclonal Mouse anti-Bovine DNA antibody (IgM) recognizing double-stranded and single-stranded DNA was obtained from US Biological (Swampscott, MA, USA)(1mg/ml). A nonspecific antibody, mouse IgM, was provided by Rockland Corp. (Gilbertsville, PA, USA). A rat arterial smooth muscle cell line (A10) was obtained from American Tissue Type Collection (Gaithersburg, MD, USA). Lipofectin and 1-Ethyl-3-(dimethylaminopropyl) carbodiimide hydrochloride(EDAC) were purchased from Sigma (St. Louis, MO, USA). N-Succinimidyl- 3-(2-pyridyldithiol)-propionate (SPDP) was obtained from Pierce Chemical (Rockfold, IL, USA). Amicon Ultra-4 Centrifugal Filter Devices with 100,000 NMWL low-binding Ultracel® membranes were obtained from Millipore Corp. (St. Louis, MO, USA). $Na^{125}I$ was purchased from Amersham Biosciences Company (England). Sephadex G-50 was purchased from Shanghai Chemical Reagent Plant (Shanghai, China). DNA polymerase I and Klenow fragment were purchased from TaKaRa Biotechnology (Dalian) Co., Ltd. $[-^{32}P]$ dATP was obtained from Furui Bioengineering Corp. (Beijing, China). GFP expression was assessed using fluorescent microscopy with a filter calibrated to detect fluorescent isothiocyanate (FITC; NIKON Inc., Melville, NY). Vectashield ®Mounting Medium for fluorescence with DAPI was purchased from Vector Laboratories, Inc. (Burlingame, VA). Mustang® 316L stainless steel coronary stents (2.5×13mm) were a gift from MicroPort Medical (Shanghai) Co., Ltd.

Protocol for Preparing pDNA –tethered Stent

The collagen coating of stents was performed as previously described *(5)*. The collagen-coated stent was reacted with SPDP at room temperature for 2 hours followed by reduction with dithiothreitol (DTT) to introduce SH- groups onto the stents. Separately, the mouse monoclonal anti-bovine DNA antibody was reacted with an excess of SPDP (without further reduction) to introduce dithiol groups on the antibody molecules. The unconjugated SPDP was removed by filtration through Amicon Ultra-4 Centrifugal Filter Devices. The dithiol-activated anti-DNA antibody was then chemically linked to the stent through the thiol-exchanging reaction. The antibody-bound stent was then incubated in a *pEGFP-N3* DNA solution (20 µg pure pDNA in 200 µl DMEM) at 37°C for 1 hour followed by an extensive rinse with copious PBS solution. The stent was further reacted with Lipofectin reagent (5 µl/200 µl) and incubated at room temperature for 35 minutes before being put into cells or implanted into rabbits' vessels. A nonspecific antibody (mouse IgM) was attached to the collagen-coated stent in the same manner as the control. In some cases, physically absorbing the antibody or pDNA onto collagen-coated stents was used as controls.

Binding Capacity and Stability of ^{125}I Labeled Anti-DNA Antibody on Collagen film

Anti-DNA antibody (IgM) was iodinated by a modified chloramine-T procedure *(40)*. Briefly, a 250µl aliquot of anti-DNA antibody (1mg/ml), 3.5µl Na^{125}I (37MBq) and 60µl chloramine-T (5mg/ml, pH 7.5) were mixed in PBS solution and incubated for 3 minutes at room temperature; the reaction was stopped by adding 80µl of a sodium metabisulphite solution (10mg/ml, pH7.5). The ^{125}I-labeled antibody was purified by gel filtration on a Sephadex G-50 column to remove un-reacted iodine. The purified ^{125}I–labeled antibody was activated with SPDP and chemically linked on to collagen film as described above (n=5). Control films were made by directly soaking collagen films in the ^{125}I-labeled antibody solution. Both types of antibody-loaded films were subjected to the same PBS rinse procedure. To ensure complete removal of unbound antibody, the eluting solution was monitored on a gamma counter until the radioactivity level reached background. The binding capacity of the antibody on collagen film was determined by measuring the radioactivity remaining on the film. To evaluate the binding stability of the ^{125}I-labeled antibody, films were incubated in PBS at 37°C with gentle shaking (140 rpm). At predetermined time point, the sample solution was replaced with same volume of fresh PBS and the radioactivity remaining on film was measured with the gamma counter.

Loading Capacity and Stability of pDNA on Anti-DNA Antibody Modified Stent

The ^{32}P-labeled pDNA was prepared by nick-translation as previously reported *(41)* with some modifications. An aliquot of 5×10^{-4}U DNase I was used to nick 2mg *pEGFP-N3*. The ^{32}P-labeled plasmid was separated from the reaction mixture by ethanol precipitation and purified using an Amicon Ultra-4 centrifugal filter device. The ^{32}P-labeled pDNA was recovered in TE buffer at a concentration of 1g/l solution. The specific radioactivity (CPM/1g) of the ^{32}P-labeled pDNA was then determined by liquid scintillation. A 10ml aliquot of PBS was chosen as solvent for liquid scintillation counting (without any other ingredients). Each antibody-modified stent was reacted with 100µg of labeled pDNA followed by a PBS rinse to remove free pDNA; rinses were carried out until the radioactivity of rinse solution returned to background levels. The radioactivity remaining on stent was counted as the total loading capacity of pDNA. To evaluate the binding stability of the tethered pDNA, the stents were incubated in PBS at 37□ with gentle shaking (140rpm). At each predetermined time point, six stents and the corresponding eluting solutions were collected for radioactivity measurement. The collagen-coated stent without anti-DNA antibody (n=5) was tested in the same way as control.

Cell Transduction by pEGFP-N3-tethered Stents inVitro

The stents were incubated in a 1×10^6 A10 cell (rat arterial smooth muscle cells) suspension for 1 hour at 37°C before being placed into 35 mm cell culture plates. A suspension of 1×10^5 A10 cells in Dulbecco's Modified Eagle Medium (DMEM) was added. The cells were incubated in serum-free medium for 5 hours followed by the addition of Fetal Bovine Serum (FBS) to a final concentration of 5%. The culture medium was changed to growth medium (DMEM+10%FBS+1% penicillin/Streptomycin) after 24 hours. The pEGFP-N3 gene expression, as evidenced by FITC-positive cells, was observed on a Nikon Eclipse E800 fluorescent microscope (NIKON) equipped with SPOT Version 3.02 software (SPOT Diagnostic Instruments, Sterling Heights, MI) and photographed after 3 days of cell culture incubation.

In Vivo Gene Expression

Eight male New Zealand White Rabbits (3.0-3.5kg) were purchased from Tianjin Medical Laboratory Animals Center (Tianjin, China). The Administrative Committee on Animal Research in the Institute of Biomedical Engineering, Chinese Academy of Medical Science approved all the protocols for animal experiments. Also, all animal experiments were performed in compliance with the Guiding Principles for the Care and Use of Laboratory Animals, Peking Union Medical College, China.

The animals were sedated with intravenous infusion of sodium pentobarbital (30 mg/kg) and the right carotid artery was exposed by surgical procedure. Two microvascular clips were used to clamp the distal and proximal segments; a small hole was then made with scissors, proximal to the distal temporary clip Each *pEGFP-N3*-tethered collagen-coated stents, premounted on a balloon catheter, was inserted visually into the proximal carotid artery utilizing

988kPa inflation (balloon: artery ration=1.2-1.3:1). The balloon was deflated for 30 seconds and withdrawn from the blood vessel. The animal was allowed to recover under general anesthesia. The control stents (*pEGFP-N3* tethered by a non-specific IgM) were also implanted into six rabbits. All the animals were given 80,000 units penicillin and 625 units heparin intravenously during procedure. The rabbits received 40 mg of aspirin orally 24 hours before surgery and each day thereafter. All the animals were sacrificed by an overdose of sodium pentobarbital 7 days after implantation. The arterial samples were retrieved and evaluated according to previous method *[5]*. Briefly, the frozen-sectioned samples were observed under a fluorescent microscope. GFP immunohistochemistry was performed and Hematocylin and Eosin-stained sections were examined under light microscopy. Samples of the lung, liver, kidney and distal vessel were harvested and examined for gene expression in biodistribution studies by PCR analyses *[6]*. For cell counting, the samples were fixed with 4% paraformaldehyde for 10 minutes and cover-slipped using Vectashield ®mounting medium. The efficiency of transduction was expressed by "transduction efficiency" which was defined as the fraction of GFP-expressing cells determined by DAPI fluorescence in each visual field. For each stent, three random fields were counted.

Statistical Methodology

Cell culture studies were carried out in replicates of five. Data for all experiments were expressed as means plus or minus the standard error of the mean (SEM). The significance of differences was assessed using Student's t–test or analysis of variance (ANOVA) and was termed significant when $p<0.05$.

Acknowledgment

These studies were financially supported by projects from Tianjin Committee of Science and Technology (043803011), the NSFC of China (50473059) and The University Doctoral Program Foundation (20030023004). We thank Yongzhe Che (Medical College of Nankai University, Tianjin, China) for his direction and efforts in animal surgery.

References

1. Sharif, F.; Daly, K.; Crowley, J.; O'Brien T. *Cardiovasc Res* **2004**, *64*, 208-216.
2. Mizuguchi, H.;, Hayakawa T. *Hum Gene Ther.* **2004**, *15,* 1034-1044

260

3. Bennett, M.R.; O'Sullivan, M. *Pharmacol Ther* **2001**, *91*, 149-166.
4. Rutanen, J.; Markkanen, J.; Yla-Herttuala, S. *Drugs* **2002**, *62*, 1575-1585.
5. Klugherz, B.D., et al. *Hum Gene Ther* **2002**, *13*, 443-454.
6. Klugherz, B.D., et al. *Nat Biotechnol* **2000**, *18*, 1181-1184.
7. Takahashi, A.; Palmer-Opolski, M.; Smith, R.C.; Walsh, K. *Gene Ther* **2003**, *10*, 1471-1478.
8. Ye, Y.W., et al. *Ann Biomed Eng* **1998**, *26*, 398-408.
9. Perlstein, I.; Connolly, J.M.; Cui, X.; Song, C.; Li, Q.; Jones, P.L.; Lu, Z.; DeFelice, S.; Klugherz, B.; Wilensky, R.; Levy, R.J. *Gene Ther* **2003**, *10*, 1420-1428.
10. Abrahams JM, Song C, DeFelice S, Grady MS, Diamond SL, Levy RJ. *Stroke* **2002**, *33*, 1376-1382.
11. Levy R.J, Song C, Tallapragada, et al. *Gene Ther.* **2001**, *8*, 659-667.
12. Stachelek S. J; Song, C.; Alferiev, I. et al. *Gene Ther.* **2004**, *11*, 15-24.
13. Walter, D. H., et al. *Circulation* **2004**, *110*, 36-45.
14. Laitinen, M., et al. *Hum Gene Ther.* **1997**, *8*, 1737-1744.
15. Turunen, M.P., et al. *Gene Ther.* **1999**, *6*, 6-11.
16. Laitinen, M., et al. *Hum. Gene Ther.* **2000**, *11*, 263-270.
17. Numaguchi, Y., et al. *Cardiovasc Res.* **2004**, *61*, 177-185.
18. Taniyama, Y., et al. *Circulation* **2002**, *105*, 1233-1239.
19. Morishita, R.; Gibbons, G.H.; Kaneda, Y.; Ogihara, T.; Dzau, V.J. *Hypertension* **1993**, *21*, 894-899.
20. Kaneda, Y., et al. *Mol. Ther.* **2002**, *6*, 219-226.
21. Tirlapur, U.K.; Konig, K. *Nature* **2002**, *418*, 290-291.
22. Gao, X.; Huang, L. *Gene Ther* **1995**, *2*, 710-722.
23. Thierry, A.R.; Lunardi-Iskandar, Y.; Bryant, J.L.; Rabinovich, P.; Gallo, R.C.; Mahan, L.C. *Proc. Natl. Acad. Sci. USA* **1995**, *92*, 9742-9746.
24. Xu, Y.; Szoka, F. C. Jr. *Biochemistry* **1996**, *35*, 5616-5623.
25. Liu, Y., et al. *Nat. Biotechnol.* **1997**, *15*, 167-173.
26. Liu, F.; Qi, H.; Huang, L.; Liu, D. *Gene Ther.* **1997**, *4*, 517-523.
27. Burton, D. Y., et al. *Hum. Gene Ther.* **2003**, *14*, 907-922.
28. Kullo, I. J.; Simari, R. D.; Schwartz, R. S. *Arterioscler. Thromb. Vasc. Biol.* **1999**, *19*, 196-207.
29. Kopp, C. W.; de Martin, R. *Curr. Vasc. Pharmacol.* **2004**, *2*, 183-189.
30. Avrameas, A.; Gasmi, L.; Buttin, G. *J. Autoimmun.* **2001**, *16*, 383-391.
31. Bauters, C.; Isner, J.M. *Prog. Cardiovasc. Dis.* **1997**, *40*, 107-116.
32. Wu, Y. X., et al. *Di Yi Jun Yi Da Xue Xue Bao* **2003**, *23*, 1263-1265.
33. Wang, K., et al. *Mol. Ther.* **2003**, *7*, 597-603.
34. Simons, M.; Edelman, E.R.; DeKeyser, J.L.; Langer, R.; Rosenberg, R.D. *Nature* **1992**, *359*, 67-70.
35. Kataoka, C. et al. *Am. J. Physiol. Heart Circ. Physiol.* **2004**, *286*, H768-774.
36. Kitamoto, S.; Egashira, K. *Expert Rev. Cardiovasc. Ther.* **2003**, *1*, 393-400.
37. Kitamoto, S.; Egashira, K.; Takeshita, A. *J. Pharmacol. Sci.* **2003**, *91*, 192-196.

38. Kipshidze, N.N., et al. *J. Am. Coll. Cardiol.* **2002**, *39*, 1686-1691.
39. Jeschke, M. G.; Klein, D. *Gene Ther.* **2004**, *11*, 847-855.
40. Hussain, A. A; Awad, R.; Crooks, P. A.; Dittert L. W. *Anal Biochem* **1993**, *214*, 495-499.
41. Wheeler, V. C.; Coutelle. C. *Anal Biochem* **1995**, *225*, 374-376.

Agricultural and Food Applications and Analytical Methodologies

Chapter 14

A Review: Controlled Release Systems for Agricultural and Food Applications

LinShu Liu[1], Joseph Kost[2], Marshall L. Fishman[1], and Kevin B. Hicks[1]

[1]Eastern Regional Research Center, Agricultural Research Service, U. S. Department of Agriculture, 600 East Mermaid Lane, Wyndmoor, PA 19038
[2]Department of Chemical Engineering, Ben-Gurion University, Beer-Sheve 84105, Israel

Controlled release systems are widely used in numerous applications. Natural polymers, such as plant polysaccharides, have been used to construct carriers of bioactive substances and deliver them in a designed manner. Polysaccharides, approaches and fabrication techniques used in controlled release systems for food preservation, food quality improvement, food packaging and fertilizer management are reviewed.

In 1970's Dr. Robert S. Langer created the first matrix, which released macromolecular substances in a controlled manner (*1, 2*). It is a great step in the history of pharmaceutical science and polymer materials. Before Langer's invention, studies and applications of controlled release technology had adhered for decades to small molecules, such as the release of mineral nutrients from encapsulate fertilizers. People had believed that only small molecules could be released from polymeric matrices. Inspired by Langer's invention, the technology has dramatically undergone great developments in the last three decades. Countless controlled release systems (CRS) have been developed,

265

which are effectively used for the administration of drugs with high molecular weights such as proteins, polypeptides, polysaccharides, growth factors, siRNA and DNA, as well as the drugs traditionally obtained by organic synthesis. Many drug carriers employed vary in chemical structures and take a continuum of physical states ranging from matrix, capsule, vesicle, liposome, chip, hydrogel, ointment, to micro- and nanospheres. Being promoted by the clinical demands and pharmaceutical industries, the uses of CRS cover almost every area of pharmacy and medicine: to control the duration and action of the drug and drug's level in the human body; to target the drug to particular organs or cells in the body; to overcome certain tissue barriers; and to simplify and improve the efficiency of disease diagnosis (3). The CRS is now growing into a $60 billion market worldwide and expected to grow 10% per year. Furthermore, theoretical and systematic studies in the science and technology of CRS further broaden its applications. Besides for biomedical uses, CRS is now applied in many areas, such as environmental protection, cosmetic, healthcare and consumer products, food industry, packaging materials and technology. CRS is undergoing a new development, from its original application of slow release of fertilizer to non-viral gene delivery tissue engineering and responsive drug delivery systems.

This article reviews the most recent developments of CRS in industries related to agrobusiness: the foods, the food packaging, and the fertilizer management.

CRS in Food Preservation and Food Quality Improvement

Polysaccharides and Food Industries

The uses of polysaccharide materials to preserve and improve food quality have a history as long as human civilization. Many recent discoveries in archaeology have revealed that the Chinese used plant exudates and plant extracts in wine preservation in 600-800 BC, the same time when Confucius and his followers were busy in developing Eastern philosophy and ethics. Similar stories also appear in Western literature. St. John's bread is a good example. As recorded in the Bible, it was locust beans that helped St. John, during his wanderings in the wilderness, to travel out of the desert to preach and baptise (Matthew III: 4). Locust bean gum was then used as an additive of breads, which are available now in many parts of the world. The bread is called St. John's bread to honor John. It is well established in much literature that polysaccharide additives can retain food flavors and release them in a controlled manner, extend shelf-life of foods, and provide the foods more appreciable texture by interacting with food components (4). The chemical and biochemical mechanisms have been thoroughly evaluated. However, how our ancestors learned to use this technique is still a mystery. We wonder at their magic hands and smart brains.

Many polysaccharides from various plants have been isolated and identified. Some of them are extracted or exuded from land plants and seaweeds; some are prepared by mechanical milling, followed by mechanical isolation. Some come from bacterial synthesis and fermentation, and some are the products of chemical modification of existing natural fibers. Interestingly, polysaccharides often begin their commercial uses as food additives even though many of them are indigestible.

Table I lists several edible polysaccharides, which are popularly used in food industries. These polysaccharides function in thickening, whipping, swelling, gelling, coating, binding, clouding, flocculating, stabilizing, and molding in many types of foods and beverages. The use of polysaccharides causes changes in food texture and nutritional quality, and alters the perceived intensities of odor, taste and flavor by masking, retaining, or controlled release technologies.

Pectin and other Polysaccharide Gels for the Controlled Release of Flavor and Aroma

As people continue looking for higher life standards, demand for healthy foods has increased. More "roughage foods" are consumed, and plant polysaccharides with physiological impact, such as pectin, have received more attention. It has been demonstrated that pectin can reduce the cholesterol content in the blood, has affects on fat metabolism without contributing in energy supply. Pectin has been applied to replace fat in foods that contain an excess of calories. Currently pectin is used in the production of bread, mayonnaise, cake, tomato ketchup, cloudy juice, ice cream, and jellied sweets. The inclusion of pectin in food products not only modifies the viscosity and texture of the foods, but also has an impact on food taste attributes, such as: flavor, sour, bitter and salty. All of these functional senses are among the higher characteristics in determining food quality when human consume foods. Since flavoring is frequently the greatest single cost component in manufactured foods, the efficiency of "flavor absorbing/releasing" is of considerable commercial importance. Food additives that can improve the stability of flavor, aroma, and the taste of the products are preferable (5-12).

It is well known in the food industry that the amount of flavoring required to produce the same subjective flavor intensity is often much higher in thickened or structured products than in more fluid systems. The factors that affect flavor release from foods are phase partition and mass transport. Flavor release from products to vapor phase will take place only if the phase equilibrium is disturbed. At equilibrium the concentrations in these phases shows the following relationship:

$$K_{ag} = C_a/C_g$$

where K_{ag} is the conventional vapor-gel partition coefficient and C_a and C_g are the concentrations of the flavor compound in air and gel phases, respectively.

Table I. Classification of Edible Hydrocolloids

Exudates	Extracts	Flours	Biosynthetic or Fermentation	Chemical Modification
	Seaweed:	Seed:		Cellulose derivatives:
Arabic	Agar	Guar	Dextran	Carboxymethycellulose
Tragacanth	Alginates	Locust bean	Xanthan	Methylcellulose
Karaya	Carrageenans		Curdlan	Hydroxyproplycellulose
Ghatti	Furcellaran	Cereal:		Hydroxyproplymethyl cellulose
	Land Plant:	Starches		Other derivatives:
	Pectin	Microcrystalline cellulose		Modified starches
	Arabinoglactan			
	Animal:			Low methoxyl pectin
	Gelatin			Propylene glycol alginate

Cited from Reference #5 with permission

Table II. Vapor/Matrix Partition Coefficient ($K \times 10^3$) for Aroma Compounds in Different Formulations: Water, Glucose Syrups, Corn Syrups and Pectin Gels

Formulations	3-Methyl-1-Butanol	2-Phenyl Ethanol	Ethyl Phenyl Glycidate
Water	47.9 ± 0.06	2.3 ± 0.06	19.6 ± 2.4
Glucose-syrup	219.0 ± 10.27	6.4 ± 0.06	33.5 ± 2.4
Corn syrup	180.0 ± 12.70	4.4 ± 0.02	24.8 ± 2.5
Pectin gel	133.0 ± 12.18	3.4 ± 0.15	16.5 ± 2.2

SOURCE: Modified with permission from reference 7. Copyright 2003.

Table II shows the values of K_{ag} of three aromas in four flavor formulations: water, glucose syrup (G syrup), corn syrup (C syrup), and sugar-pectin gel. The K_{ag} value in water is significantly lower than those obtained in the presence of sugars and pectin. Sugars interact with water, resulting in a "salting-out" effect on the incorporated aroma, by which the concentration of aroma vapor in the air phase increases. Corn syrup, a hydrolytic product of corn starch, is a mixture of macromolecules with different molecular weights; whereas, glucose is a small molecule (MW of 180). At same concentration, syrup containing glucose has more binding sites for water than that containing hydrolyzed corn starch. Therefore, there is more free water available for flavor molecules to dissolve in C syrup. Besides water activity, physical barrier and possible chemical interactions are two major concerns in flavor release from food products. The addition of pectin causes a further decrease in the partition coefficients of the three aromas in comparison with C syrup. This could be attributed to the formation of diffusion barrier in pectin gel and may also relate to the interactions between the aroma and pectin. It is generally accepted that the diffusion of a solute is inversely proportional to viscosity. Since only free dissolved molecules exert a vapor pressure, the binding of flavor to matrix or complex formation between flavor and matrix will suppress flavor release. The chemical inhibition of flavor release may not be preferable in some cases (concerns related to the changes in structure and function of flavor molecules). High molecular weight sugar chains entangle to form gel networks, which is responsible for retarding the diffusion of flavor molecules. Thus, gel texture plays a role in determining perceived flavor intensity. Many studies have been done in the attempt to correlate the rheological properties of polysaccharides in solution with the structures of their gels and the release of incorporated flavors (6, 7, 9-14).

Another well-used plant polysaccharide is guar gum. Guar gum is a neutral polymer; extensively used in thickened products. In solutions below critical concentration, guar behaves as extended "random coil". The effective hydrodynamic volume of each molecule is independent of overall concentration, facilitating molecular interpenetration of the bulk polymers. In contrast to non-charged polysaccharides, charged polymers such as alginate, carrageenan and sodium carboxymethylcellulose, display two reactions with water simultaneously: the reaction of water with macromolecules and the reactions of water with the accompanied counter ions. When used in water at different concentrations, the volume occupied by the individual polymer chains is not constant; it decreases with increasing polymer concentration in response to the associated increases in ionic strength from the counter ions to the polymer.

Encapsulation of Flavor/Aroma and other Food Components

A new strategy to regulate flavor release from food products is to encapsulate flavors or aromas into carrier matrices, which are usually within the size of micro- or nano- scale. The flavor loaded delivery vehicles are then incorporated into food products. The encapsulation prolongs the flavors' shelf time and enhances the tolerance of flavor/aroma to the variation of processing and storage temperature. Furthermore, the controlled release of food ingredients from capsules improves the effectiveness of food additives; broadens the application range of food ingredients with optimal dosage. Many sophisticated materials and technologies, which were once developed for pharmaceutical applications, are now used widely in food industries. The most impressive is the use of liposomes. Liposome encapsulation that can be produced continuously in large scale without organic solvent and ultrasonic irradiation now is popular in food industries as a routing process (15, 16). It has been well established that liposomes enhance the stability of water soluble materials in high water activity environments. By altering the shell materials, liposomes can specifically deliver their contents to a certain part of the food stuff. For example, liposome-encapsulated flavor-producing enzymes concentrate preferably in the curd during cheese formation, while non-encapsulated enzymes are evenly distributed in the whole milk (17, 18).

Alginate/calcium coacervation is another method that is extensively used in encapsulation. Due to its macro-porous structure, the method is more suitable for substances with extremely large molecular weight and size, such as enzymes. A recently developed emulsion-based process has made scaling-up of alginate/calcium chloride coacervates available. By the method, an aqueous solution of alginate and active ingredient is added to a non-aqueous phase followed by homogenization to form micro-droplets. The emulsified droplets are then cross-linked by the addition of calcium chloride solution (19, 20).

Cyclodextrin, a cyclic molecule made up of glucose monomers, is produced from starch by enzymatic process using cyclodextrin glycosyltransferase. The seven-unit β-cyclodextrin is the most studied cyclodextrin for encapsulation. Characterization of cyclodextrin and its use in food industries have been extensively conducted (21, 22). Small organic molecules such as artificial flavors can displace the water in the inner cavity (5-8 Å) and stay within the cavity to form thermodynamically stable complexes. The flavor compounds release in the presence of a better substances (size, shape, charge, interaction with the inner cavity, etc), which displace the flavor from the cavity. However, any food components significantly larger than the "cavity" will not "fit" in, which leads to lower encapsulation efficiency and consequently, a complicated purification process.

Other encapsulation methods used in food industries include spray drying (*23*), spray cooling and spray chilling (*24*), extrusion (*25, 26*), coacervation (*27, 28*), centrifugal co-extrusion (*29*), spinning disk (*30*), and fluidized bed (*31, 32*) technologies. New encapsulation technology and shell materials are continually devised and invented. Although in comparison with the wide range of encapsulated products that have been marketed in the pharmaceutical and cosmetic industries, encapsulation only occupies a small portion in food industries. There is a clear trend that CRS applications in food technologies and industries are increasing. In 2002, more than 1000 U.S. Patent Applications concerning encapsulation technology and its applications were filed, among them 1/3 related to food industries. This ratio was 1/10 in 1992.

For use in foods, the matrix materials should be ingestible and able to address specific requirements generated by the different environments passing from the mouth to the gut, such as saliva digestion, effect of mastication, and the variation of pH, proteases and microflora. Many natural polymers have been used for this purpose. The approach to encapsulate flavor/aroma in edible micro- or nano- spheres to release them in a controlled manner has attracted increasing attention and resources in its research and development.

Flavor loss or flavor degradation usually occurs during processing or storage. As a complement, flavor releasing films were developed. The films can be lined on the inner wall of food cans, which fill the headspace with fragrance to please consumers when the package is first opened.

CRS in Food Packaging Materials

Active Packaging

There is an increasing demand for various packaging materials. The global packaging market in 2004 was about 300 billion dollars while food packaging is the largest sector in the packaging industry about $60 billion in 2004 and is expected to grow to $74 billion by 2008 (*33, 34*). Advanced food packaging not only minimizes food loss and protects foods from contamination, it is also designed to satisfy consumer trends for high quality foods, fresh-like products, and convenient or ready-to-eat foods. New concepts and technologies have been proposed and/or developed to produce such packaging materials, which are defined as active packaging or named as intelligent packaging or smart packaging (*33, 35*). In contrast to passive packaging that offers physical barriers to protect contents from environmental influences, active packaging provides chemical and biological means to moderate changes occurred on the surfaces and in the headspaces of packaged materials. By controlling microbial growth on meats and ripening of vegetables and fruits, active packaging is able to extend the shelf time of foods and improve their quality. The technologies of active packaging include the use of oxygen and moisture scavengers and carbon dioxide emitters, which retard or suppress microbial growth; the spreading and spraying of ethanol, which inhibits mold growth; the use of ethylene absorbers, which reduce the respiration rate of climacteric crops; the application of controlled delivery systems for flavor or antimicrobial release, and a variety of indicators of temperature, humidity and other environmental factors. For detailed information, readers can refer to several recent review papers (*35-37*). In addition, many active packaging materials are made from biodegradable polymers which are friendly to the environment and thus limit problems related to pollution and disposal. In this section, we focus on the applications of CRS in active food packaging.

Controlled Release of Antimicrobials

Although the application of CRS in active packaging is still in its early stage, the development is rapid and very promising (*33-38*). The recent developments include the controlled release of antimicrobial agents, the preparation of antimicrobial surfaces, and the on-site generation of free radicals. A large number of antimicrobials have been used in food preservation (Table III). Traditionally, antimicrobials were directly added into food bulks by dipping the food in antimicrobial solutions. Therefore, a large amount of antimicrobial is needed because the antimicrobial may be diluted by migration into the food matrix or inactivated by food components. By the use of controlled release technology, antimicrobial agents are incorporated on the surfaces of packaging materials by laminating, coating and adsorption; or within the materials by embedding, mixing, or dispersing. The incorporated substances are, then, allowed to gradually release out at desired rates into the foods or the environment surrounding the foods and exercise their functions there. Alternatively, the active compounds can be retained on the surfaces of packaging materials by ionic or covalent immobilization and act upon the contact with targeted microorganism. By these new strategies, only a small amount of antimicrobial is required.

Antimicrobial agents that are used in active packaging can be sorted into five types: organic acids, enzymes, bacteriocins, fungicides, and plant extracts. They differ from each other in function, method of preparation, mechanism and objectives. The first type, antimicrobial organic compounds, include lactic acid,

Table III. Application of Antimicrobial Packaging in Different Food Systems

Food Product	Antimicrobial Agent	Target Microorganism
Meat, fish, and poultry		
Beef	Pediocin	Listeria monocytogenes
	Nisin	Brochothrix thermosphacta
	Triclosan	Brochothrix thermosphacta/Salmonella typhimurium/Escherichia coli H157:H7/Bacillus subtilis
Ground beef	Grapefruit seed extract	Micrococcus flavus/E. coli/ Staphylococcus aureus/B. subtilis
Ham	Lacticin 3147 and nisin	Lactococcus lactis subsp. lactis/Listeria innocua
		Staphylococcus aureus
Poultry	Nisin	Salmonella typhimurium
Ham/bologna/pastrami	Acetic acid/ propionic acid	Lactobacillus askei/Serratia liquefaciens
Vacuum packaged beef carcass	Nisin	Lactobacillus helveticus/Brohothrix thermosphacta
Fresh broiler skin	Nisin	Salmonella typhimurium
Vegetable type products		
Strawberry	Potassium sorbate/citric acid	Aerobic mesophilic/psychrotrophic/ molds/yeast/coliforms
Tomato	Citric acid/acetic acid/sorbic acid/ethanol	Salmonella montevideo
Lettuce/soybean sprouts	Grapefruit seed extract	E. coli/Staphylococcus aureus
Milk and dairy products		
Skim milk	Nisin	Lactobacillus curvatus
Cheddar cheese	Lacticin 3147 and nisin	Lactococcus lactis subsp. Lactis/ Listeria innocua/Staphylococcus aureus
Cheese	Imazalil	Penicillium sp./Aspergillus toxicarius

SOURCE: Reproduced with permission from reference 19. Copyright 2000.

benzoic acid, sorbic acid, etc, and their corresponding acid anhydrides. All these organic acids show higher antimicrobial activity at lower solution pH than at higher ones. At the lower solution pH, organic acids are most likely in their protonated form. The inhibitory mechanism is considered to include penetration of organic acid compounds across plasma membrane in their undissociated state. The second type of antimicrobial enzymes includes lysozyme, chitinase, and glycose oxidase, etc. Lysozyme is a single chain peptide with a specific hydrolytic activity against both gram-positive and gram-negative bacteria. Lysozyme attacks the β-1,4-glycosidic linkage of N-acetylmuramic acid and N-aceylglucosamine on the cell walls of bacterial, resulting in the damage of cell wall integrity and the subsequent lysis of bacterial cells. For the third type of antimicrobial, the best studied and frequently used bacteriocin is nisin. Nisin is a polypeptide produced by the lactic acid bacterium, *Lactococcus lactis*, with effective antimicrobial activity against gram-positive organism. Imazalil and benomyl are among the fourth type of antimicrobial agent, fungicides that have been permitted to be used on fresh fruits and cheese to prevent mold infection. The fifth type, plants and vegetables extracts, is receiving increasing interest and attention in active packaging. Plant extracts include grape fruit seed extracts and essential oils, which express a wide spectrum of microbial growth inhibition activities. The antimicrobial properties of essential oils are mainly attributed to the phenolic components. The active components in the grape fruit seed extracts are naringin, ascorbic acid and hesperidins. As antimicrobial agents are used to prevent food spoilage, they are generally stable within a specified storage time, and are not likely to decompose under food processing conditions. Furthermore, in order to be active, the antimicrobial agents should contact the foods directly, thus ideal antimicrobial compounds used in active packaging must be food-grade. Nevertheless, none of these agents has a complete antimicrobial spectrum to act against bacteria, molds and yeasts.

Controlled Release of Volatile Antimicrobials

Another approach is the delivery of volatile antimicrobials that can penetrate the food matrix. The application of controlled release of antimicrobial volatiles in active packaging avoids the direct contact of packaging materials with food products. That is especially suitable for packaging of ground meat or fresh cut products. Several delivery systems for the controlled release of volatile antimicrobials have been developed. Polymeric devices containing sodium chlorite and acid precursors release chlorine dioxide on response to moisture. Pads of sodium metabisulfite release sulfur dioxide in a sustained manner. Other delivery systems that release carbon dioxide, chlorinated phenoxy compounds, ethanol vapor, or allylisothiocyanate have been commercialized. Volatile constituents that are extracted from plants such as Japanese horseradish and mustard oil are also employed.

Release of Free Radicals

The third approach is to control the generation and release of free radicals on site to suppress bacterial growth. The presence of oxygen in food packages may facilitate microbial growth and off-odors development. The control of oxygen level in food packages is important to maintain food quality. A system has been developed to convert oxygen to peroxide radicals that inhibit growth of microorganisms. Several enzymes and "electron-pool" (usually, metal-substituted porous ceramics) are used for this purpose. Glucose oxidase can be immobilized on the surfaces of packaging materials. The immobilized enzymes interact with oxygen and glucose, a food component, yielding antimicrobial hydrogen peroxide. With the same objective, glucose oxidase and β-galactosidase can be co-immobilized on the inner surface of milk package to produce hydrogen peroxide, which, in turn, activate lactoperoxidase. Peroxide radicals also can be produced by the use of silver-substituted ceramics such as zirconium phosphate ceramics ($Ag_xH_{1-x}Zr_2(PO_4)_3$) and zeolite. When the ceramics are exposed to white light, oxygen molecules adsorbed on the structures will be reduced by silver to $O_2^-\bullet$ radicals:

$$Zr_2(PO_4)_3, + h\nu \rightarrow Zr_2(PO_4)_3, + e^-(\text{surfaces-Ag}^+)\text{-}O_2 + e^- \rightarrow (\text{surfaces-Ag}^0)\text{-}O_2 \rightarrow$$
$$(\text{surfaces-Ag}^+) + O_2^-\bullet$$

Without light irradiation, silver-substituted materials act against bacterial and mold by the release of silver ions from the structures. Because aqueous solutions from the food enter the porous structure and wet down the cavities, bacterial and mold uptake the silver ions. The bound silver ions disrupt the activity of bacterial cells by replacing other essential metals. Silver-substituted compounds can be incorporate into food contact polymers by lamination or coating.

Biomass Materials used for Active Packaging

A number of plant-derived polysaccharides have been used as matrix materials in antimicrobial packaging. These include starch, pectin, carageenans, cellulose and cellulose derivatives; also included are polymers that can be synthesized from plant-polysaccharide-derived monomers, such as poly(lactic acid). Polymers, which are produced by microorganisms or extracted from marine products such as xanthan, curdlan, dextran, pullulan, chitin, chitosan, and alginate, having the structures and properties similar to those plant-derived, are often discussed together with plant-derived polysaccharides in some literatures (*35, 37*). Animal byproducts originating materials, such as collagen and gelatin, are also used in food packaging. In comparison with petroleum-derived thermoplastics, biobased materials are biodegradable and some are food-grade.

However, they are always mechanically weak and less water resistant than petrochemically derived ones. Thermal properties of biomass-based materials are another concern in processing and engineering packaging materials. Other obstacles for applying natural polymers in packaging materials are: (1) the incompatibility between some biobased materials and certain antimicrobial components; (2) activity loss of antimicrobial agents when they are incorporated to, immobilized on, or released from packaging materials; (3) the heat lability of antimicrobial compounds during extrusion of antimicrobials together with biobased materials.

Nanotechnologies and Nanostructures in Active Packaging

The introduction of nanotechnology will undoubtedly change the food industry. It has been expected that the most promising application of nanotechnology in food industry will be in active packaging. On the other hand, studies on nanotechnology in this field are currently at zero commercial activity and will be limited to high-value products over the next few years. Concerns from packaging industry are: (1) when will application of nanotechnology be commercially available? (2) will this technology be cost effective? We believe that these are the topics needed to be addressed.

CRS in Fertilizer Management

CRS-Derived Fertilizers in Farm and Non-Farm Applications

The agrochemical industry produces large volumes of various fertilizers. In 1999, the worldwide usage of nitrogen (N), phosphorous (P) and potassium (K) was 81×10^6, 14×10^6 and 18×10^6 tons, respectively. These numbers increase at 2% annually. The application of fertilizers plays an increasing important role in crop supply. By 2020, more than 70% of grain yield will have to depend on the effectiveness of fertilizers (39).

The use of fertilizers has an impact on health and environmental aspects. About 98% of fertilizers are now applied conveniently on the top of the soil in water-soluble forms (40). They are easily removed from the soil by rain or irrigation water. It has been estimated that the loss of water-soluble fertilizer could be up to 30% or 50%, depending on the weather and soil conditions. Therefore, fertilizers are sometimes used in the amounts greatly exceeding those required to ensure an adequate level in plants for a suitable period. Poor fertilization management has led to risks of surface or ground water pollution and raised serious health concerns. It has been reported that in many places throughout the world, N and P accumulated at levels greatly exceeding the number that stable ecosystems can accommodate. The higher N and P accumulation is a threat for human health. For example, high nitrate levels in ecosystems are reported to be associated with birth defects, i.e., methemoglobinemia in infants (41-43) and ruminant (44, 45), gastric cancer, goiter and heart diseases (46, 47). The major environmental problem related to fertilization is the accelerated eutrophication of surface water (43, 44, 48-50) that stimulates the growth of algal biomass, which in turn, leads to the depletion of oxygen, damage of aquatic product industries, development of bad odors and initiation of aesthetic problems. Furthermore, poor fertilization management causes air pollution and soil deterioration (51-53).

In confronting these problems, efforts and resources have been placed for the improvement of nutrient supply to increase nutrient use efficiency and reduce environmental and health threats. Numerous technologies have been developed since 1960 (54, 55); among them CRS is considered the most suitable strategy. For more information on this issue, readers are strongly advised to refer to Ref. #50.

The ideal CRS-derived fertilizers are expected "to synchronize nutrient release with the pattern of plant uptake" (50). Unfortunately, this is still far from realty. Some difficulties, such as the diversities of plant, variations in climate and soil, handling conditions, and the complex interactions among plant roots, soil microorganisms and chemicals, make it impossible to tailor an ideal delivery system, which can release nutrients at the rate matching that plants uptake. Most CRS-derived commercially available fertilizers are "slow-release" fertilizers, which release nutrients in a slower manner than convenient fertilizers; only a few CRS-derived fertilizers, with varying degree of success, are able to control the rate, pattern and duration of release.

Besides the contribution to improving fertilization efficiency and environmental protection, other advantages of using CRS-derived fertilizers include the reduction of toxicity, labor, time and energy saving by lowering frequency of application. The toxicity of conventional soluble fertilizers is caused by high ionic concentrations resulted from quick dissolution; it is especially harmful for seeds. Nevertheless, the use of CRS-derived fertilizers only occupy a small portion, <2%, of the total fertilizer applications. CRS-derived fertilizers are mainly consumed by a few developed countries such as the United States, Canada, Japan, and in Europe. CRS-derived fertilizers seem more marketable for "high-cash" vegetable growth (melon, citrus and strawberry, etc), non-farm purposes (lawns and golf yards, etc), and more suitable for poor-textured soils. The prices of CRS-derived fertilizers are usually 3-10 times higher than the relative conventional fertilizers. This can be attributed to the additional processing steps and materials used in the manufacture. The higher price is the biggest obstacle to extend the uses of CRS-derived fertilizers.

Types of CRS-Derived Fertilizers and Release Mechanisms

Fertilizers, which release nutrients at a slower rate or in a controlled manner, can be classified as three categories. The first is slightly soluble fertilizers. It includes chemically modified products, such as urea-aldehyde condensation products (such as urea-formaldehyde, urea-isobutyraldehyde, and urea-acetyaldehyde/crotonaldehyde); chemically decomposing compounds, such as isobutyledene-diurea, oxamide and melamine; and inorganic low-solubility compounds (such as magnesium ammonium phosphate). Other CRS-derived products are coated fertilizers. Coated fertilizers are physically prepared by coating granules of conventional fertilizers with hydrophobic materials that control the release rate. The coating materials can be organic or inorganic polymers. Examples include thermoplastic, resin, sulfur or mineral-based coatings. Matrix-type fertilizers constitute the third major category of the CRS-derived products. The matrix-type fertilizers consist of a matrix phase through which active agents are dispersed and diffused. The materials used for matrix preparation include synthetic and natural polymers, such as polyolefins, rubber, starch, cellulose derivatives and wax. The matrix-type fertilizers have seldom been studied and are less in practice. Although CRS-derived fertilizers possess various advantages, disadvantages for certain products are obvious and therefore, improvements are required. For example, for urea-aldehyde low solubility fertilizers, a portion of nitrogen may be released at an extremely slow rate or not released out at all. With sulfur-coated fertilizers, the repeated use may lower the soil pH, and often display an initial rupture. Some coating materials, such as non-degradable thermoplastic polymers may cause environmental problems, since they accumulate in the fields.

Intensive works have been done to elucidate the mechanism of fertilizers release from CRS. Models of empirical, semiempirical and mechanistic models have been suggested for this purpose. (50, 56-63). Most of the works focus on the effect of the products structure on the release of nutrients from the fertilizers, such as coating materials, porosity, ratios between core and coating, solubility of coating materials and core materials, etc.

For coated fertilizers, a multi-stage diffusion model has been proposed to describe the nutrient release. As shown in Figure 1, following application, the granules start to absorb moisture, which dissolves the nutrients inside. Then, osmotic pressure is built-up and the granules swell. At the last stage, the dissolved nutrients are either gradually released by diffusion or immediately released if the rupture occurs. By selecting a temperature-sensitive hydrogel as the coating material, the nutrient release can be controlled by soil temperature solely, excluding the effects of soil type, water content, environmental pH and microbial activity.

Other mechanisms have been also proposed, such as erosion for coated and matrix-type fertilizers, decomposition, chemical interaction and enzymatic degradation for all types (50). However, there is still a lack of correlation

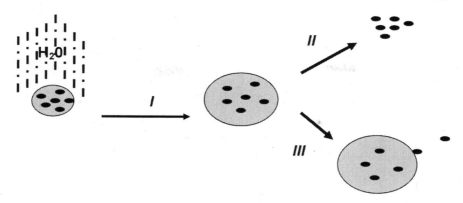

Figure 1. Scheme of the stages of the release from a polymer-coated granule: (1) water penetration, (2) dissolution, and (3) pressure build-up and/or swelling. In the last stage there are two possibilities: "failure" release if rupture occurs or "diffusion" release. (Reproduced with permission from reference 50. Copyright 2000.)

between laboratory tests and field conditions, thus, no standardized methods for reliable determination of the nutrient release pattern available as yet.

Challenges

Considerable results have been obtained in the CRS applications in the areas of food preservation and quality improvement, food packaging, and fertilizer management. With continuing progress in CRS technology, advances in science and engineering of environment, biology, chemistry and materials, we expect greater achievements in various aspects in agrobusiness. Some of the examples are: (1) A fundamental understanding of the influences of food texture on aroma perception, so that the food industry no longer needs to rely on empirical formulations. (2) Smart packaging materials that react in response to environmental stimuli. (3) Food safety nanodevices that will promptly detect harmful contaminants and can be easily used in packaging materials. (4) New procedures and strategies to lower the cost of CRS-derived fertilizer.

References

1.   Langer, R. *Chem. Eng. Commun.* **1980**, *6*, 1.
2.   Langer, R.; Peppas, N. A. *Biomaterials* **1981**, *2*, 195.

3. Langer, R.; Peppas N. *A.AIChE Journal* **2003**, *49*(12), 2990.
4. Yalpani, M., Ed.; *Industrial Polysaccharides;* Elsevier: New York, 1987.
5. Glicksman, M. *Food Hydrocolloids I;* CRC Press: New York, NY, 1989.
6. Hansson, A.; Leufven, A.; Ruth, S. V. *J. Agric. Food Chem.* **2003**, *51*, 2000.
7. Lubbers, S.; Guichard, E. *Food Chem.* **2003**, *81*, 169.
8. Rega, B.; Guichard, E.; Voilley, A. *Sci. Aliments* **2002**, *22*, 235.
9. Braudo, E. E.; Plashchina, I. G.; Kobak, V. V.; Golovnya, R. V.; Zhuravleva, I. L.; Krikunova, N. I. *Nahrung* **2000**, *44*, 173.
10. Fu, J. T.; Rao, M. A. *Food Hydrocolloids* **2001**, *15*, 93.
11. Evageliou, V.; Richardson, R. K.; Morris, E. R. *Carbohydr. Polym.* **2000**, *42*, 245.
12. Davidson, J. M.; Linforth, R. S. T.; Hollowood, T. A.; Taylor, A. J. *J. Agric. Food Chem.* **1999**, *47*, 4336.
13. Hansson, A.; Leufven, A.; Pehrson, K.; Stenlof, B. *J. Agric. Food Chem.* **2002**, *50*, 3803.
14. Stading, M.; Langton, M.; Hermansson, A.-M. *Food Hydrocolloids* **1993**, *7*, 195.
15. Frederiksen, L.; Anton, K.; van Hoogevest, P.; Keller, H. R.; Leuenberger, H. *J. Pharmaceutical Science* **1997**, *86*, 921.
16. Gouin, S. *Trends in Food Science & Technology* **2004**, *15*, 330.
17. Fresta, M.; Puglisi, G. *Microspheres, Microcapsules and Liposomes* **1999**, *2*, 639.
18. Vuillemard, J. C. *J. Microencapsulation* **2000**, *8*, 547.
19. Mofidi, N.; Aghai-Moghadam, M.; Sabolouki, M. N.; *Process Biochemistry* **2000**, *35*, 885.
20. Champagne, C. P.; Blahuta, N.; Brion, F.; Gagnon, C. *Biotechnology and Bioengineering* **2000**, *68*, 681.
21. Hedges, A.; McBrde, C. *Cereal Foods World* **1999**, *44*, 700.
22. Reineccius, T. A.; Reineccius, T. A.; Peppard, T. L. *J Food Science* **2002**, *67*, 3271.
23. Takeuchi, H.; Yasuji, T.; Yamamoto, H. *Pharmaceutical Research* **2002**, *17*, 94.
24. Morgan, R.; Blagdon, P. A. *U.S. Patent 5,204,029.* **1989**.
25. Benczedi, D.; Bouquerand, P. E. *PCT WO 01/17372 A1*, **2001**.
26. Quellet, C.; Taschi, M.; Ubbink, J. B. *U.S. Patent 0008635*, **2001**.
27. Dobetti, L.; Pantaleo, V. *J. Microencapsilation* **2002**, *19*, 139.
28. Korus, J.; Tomasik, P.; Li, C. Y. *J. Microencapsulation* **2003**, *20*, 47.
29. Sparks, R. E.; Jacobs, I. C.; Mason, N. S. *Drug Manufacturer, Technical Series,* **1999**, *3*, 177.
30. Gouin, S. *Industries Alimentaires et Agricoles* **2002**, *117*, 29.
31. Dewettinck, K.; Huyghebaert, A. *Proceedings of the 8th int Congress on Engineering and Food* **2001**, 982.
32. Matsuda, S.; Hatano, H.; Kuramoto, K.; Tsutsumi, A. *J. Chemical Engineering of Japan* **2001**, *34*, 121.

33. Cutter, C. N. *Meat Science* **2006**, *74*, 131.
34. Ozdemir, M.; Floros, *Critical Reviews in Food Science and Nutrition* **2004**, *44*, 185.
35. Tharanathan, R. N. *Trends in Food Science & Technology* **2003**, *14*, 71.
36. Vermeriren, L.; Devlieghere, F.; Debevere, J. *Food Additives and Contaminants* **2002**, *12*, 163.
37. Cha, D. S.; Chinnan, M. S. *Critical Reviews in Food Science and Nutrition* **2004**, *44*, 223.
38. Labuza, T. P.; Breebe, W. M. *J. Food Proc. Preserv.* **1989**, *13*, 1.
39. IFA. *World Fertilizer Consumption Statistics, Annual* **1999**.
40. Bolan, N. S.; Rajan, S.S.S. *Fertilizer Research* **1993**, *35*, v.
41. Supper, M.; De V. Heese, H.; Mackenzie, D.; Dempster, W. V.; Du Plessis, J.; Ferreira, J. J. *Water Res.* **1981**, *15*, 126.
42. Wilson, F. N.; Chem, C. *"Slow Release: Trye or False? A Case Study for Control"*. Fertiliser Society of London, **1988**.
43. Newbould, P. *Plant. Soil.* **1989**, *115*, 297.
44. Bockman, O. C.; Kaarstad, O.; Lie, O. H.; Richards, I. *Agriculture and Fertilizers* Agricultural Group, Norsk Hydro, Oslo. **1990**.
45. Smith, J. E.; Beutler, E. *Am. J. Public Health* **1966**, *62*, 1045.
46. Forman, D. *Cancer Survey* **1989**, *8*, 443.
47. Black, C. A. *Comments from CAST 1989-1;* Ames, Iowa. **1989**.
48. Kaap, J. D. *Proc. Agricultural Impacts on Ground Water* Omaha, NE. **1987**, 412.
49. Sims, J. T. *Commun. Soil Sci. Plant Anal.* **1998**, *29*, 1471.
50. Shaviv, A. *Advances in Agronomy* **2000**, *20(42)*, 1.
51. Keeney, D. *Third Int. Dahlia Greidinger Sym. on Fertilization and the Environment.* Haifa, Israel, **1997**.
52. Aber, J. D. *Trends Ecol. Evol.* **1992**, *7*, 220.
53. Sposito, G. *The Chemistry of Soil;* Oxford Univ. Press: New York, 1989.
54. Oertli, J. J.; Lunt, O. R. *Soil Sci. Soc. Am. Proc.* **1962**, *26(6)*, 579.
55. Lunt, O. R.; Oertli, J. J. *Soil Sci. Soc. Am. Proc.* **1962**, *26(6)*, 584.
56. Al-Zahrani, S. M. *Ind. Eng. Chem. Res.* **2000**, *39*, 367.
57. Jarosiewicz, A.; Tomaszewska, M. *J. Agric. Food Chem.* **2003**, *51*, 413.
58. Novillo, J.; Rico, M. I.; Alvarez, J. M. *J. Agric. Food Chem.* **2001**, *491*, 1298.
59. Jarosiewicz, A.; Tomaszewska, M. *J. Agric. Food Chem.* **2002**, *50*, 4634.
60. Shaviv, A.; Raban, S.; Zaidel, E. *Environ. Sci. Technol.* **2003**, *37*, 2251.
61. Chatzoudis, G. K.; Rigas, F. *J. Agric. Food Chem.* **1998**, *461*, 2830.
62. Raban, S.; Shaviv, A. *Proc. 22nd Int. Symp. Control Rel. Bioact. Mat.* CRS, Inc., **1995**, 105.
63. Fan, L. T.; Singh, S. K. *Controlled Release – A Quantitative Treatment*; Springer-Verlag, Berlin, 1990.

Chapter 15

State of the Art Mass Spectrometric and Chromatographic Techniques for Drug Analysis

Katerina Mastovska

Eastern Regional Research Center, Agricultural Research Service, U.S. Department of Agriculture, 600 East Mermaid Lane, Wyndmoor, PA 19038

In recent years, mass spectrometric (MS) and chromatographic instrumentation and techniques have scored dramatic developments, resulting in the introduction of many useful tools for analysis of both small and large drug molecules. This chapter describes state of the art MS and chromatographic techniques that can be used in the analysis of drugs in various applications, including characterization of controlled-release drug delivery systems. In applications that require minimum sample preparation, direct MS detection of drug molecules can be performed using novel atmospheric pressure ionization techniques, such as desorption electrospray ionization or direct analysis in real time. Visual information about distribution of drugs in various materials or tissues can be obtained through imaging MS, mainly using secondary ion MS or matrix assisted laser desorption/ionization. In liquid and gas chromatography combined with MS (LC- and GC-MS), drug analysis can be speeded up using various fast chromatographic techniques that are becoming practical due to the introduction of modern LC, GC, and MS instruments.

Mass spectrometry (MS) has become the analytical technique of choice in modern laboratories performing drug analysis. In complex samples, such as biological fluids or tissues, the combination of MS with a chromatographic separation provides enhanced selectivity, which is invaluable for detection of trace level components. In recent years, MS and chromatographic instrumentation and techniques have scored dramatic developments, resulting in the introduction of many useful tools for analysis of both small and large drug molecules. This chapter describes state of the art MS and chromatographic techniques that can be used in the analysis of drugs in various applications, including characterization of controlled-release drug delivery systems. The chapter discusses mainly novel approaches that require minimum sample preparation, provide high speed/high throughput analysis, and/or improve the qualitative or quantitative aspects of the analytical process.

Mass Spectrometry without Chromatographic Separation

The basic MS process involves ionization of sample molecules, followed by separation of ionized molecules and their fragments based on their mass-to-charge ratio (m/z) and detection (counting) of the ions, which results in a mass spectrum (a snapshot of ion intensities plotted against their m/z) showing the mass distribution of the ions produced from the sample. MS offers both qualitative and quantitative information; it can detect compounds, elucidate their structures, and determine their concentrations.

Direct MS measurements enable rapid analysis, especially when the analytes can be sampled directly on the sample surface at atmospheric pressure, which can be done using some recently introduced ionization techniques, such as desorption electrospray ionization (DESI) or direct analysis in real time (DART). To compensate for the lack of compound separation prior to the MS step in complex samples, MS without a chromatographic separation usually employs a highly selective MS technique, such as high-resolution, accurate mass, and/or mass-selective fragmentation (multi-stage MS^n; usually a tandem MS, MS/MS) measurements. The selectivity of MS analysis can be also greatly enhanced by using a new technology called high-field asymmetric waveform ion mobility spectrometry (FAIMS) that can separate ions prior to their introduction into the vacuum chamber of an MS instrument.

Recent developments in ionization techniques, mass analyzers, and also data processing technology have opened doors to applications that can advance a wide array of scientific disciplines. For instance, imaging MS can provide visual information about distribution of compounds in various materials, such as drugs in tissues or delivery systems, which is invaluable to scientists performing medical, material and/or drug development research.

Novel Ionization Techniques for Direct Sampling of Drugs on Surfaces

In the history of MS development, the most significant breakthrough in drug analysis can be probably attributed to the introduction of two ionization techniques in the late 1980s: (i) matrix assisted laser desorption ionization (MALDI), which is applicable to the analysis of solid materials *(1)* and (ii) electrospray ionization (ESI), which is used for the analysis of solutions *(2)*. This breakthrough was well recognized in 2002, when John B. Fenn and Koichi Tanaka received the Nobel Prize in chemistry for their work on ESI and MALDI, respectively. The introduction of ESI started the development of modern liquid chromatography-mass spectrometry (LC-MS) instrumentation, which revolutionized the analysis of polar compounds and became an indispensable tool in the analysis of small and large drug molecules.

MALDI belongs to a family of desorption ionization techniques, which desorb and ionize molecules from a surface of a condensed-phase sample by impacting it with projectiles (see Figure 1), such as photons in laser desorption (including MALDI), translationally excited atoms (in fast atom bombardment, FAB), or energetic ions (in secondary ion mass spectrometry, SIMS). MALDI requires the sample to be mixed with an excessive amount of a UV-absorbing matrix compound, which is ionized by the laser and then ionizes sample molecules by proton transfer.

- photons (MALDI, APMALDI)
- excited atoms (FAB)
- high-energy ions (SIMS)
- ions and charged droplets (DESI)
- excited gas molecules (DART)

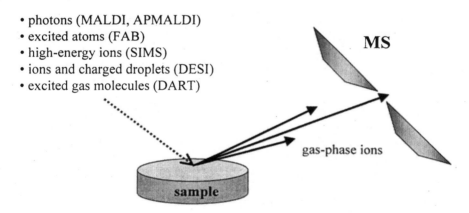

Figure 1. Illustration of the process involved in MS ionization methods for direct sampling on surfaces of condensed-phase samples in vacuum (MALDI, FAB, SIMS) or at atmospheric pressure (APMALDI, DESI, DART). See the text for the explanation of the acronyms.

The desorption/ionization process in MALDI, FAB, or SIMS takes place in vacuum, thus the sample is not accessible during the analysis. Certain applications, especially those in the bioanalytical field, would benefit from direct sampling at ambient conditions, which would allow full access to the sample for observation and additional physical processing during the analysis.

Atmospheric pressure MALDI (APMALDI), first described in 2000 *(3, 4)*, was the first ionization technique in which a condensed-phase sample could be examined at atmospheric pressure. However, APMALDI still requires sample dilution/coating with a UV-absorbing matrix compound. Furthermore, the source must be enclosed to protect the operator from potential exposure to laser radiation. Recently introduced ionization methods DESI *(5)* and DART *(6)* (reported in 2004 and 2005, respectively) do not pose these limitations. They require essentially no sample preparation and allow full access to the sample while mass spectra are being recorded.

Desorption Electrospray Ionization (DESI)

DESI is a soft ionization technique that uses gas-phase solvent ions and charged microdroplets of electrosprayed solvent interacting with a condensed phase sample to yield gas-phase ions under ambient conditions *(5, 7)*. Several possible ionization mechanisms have been suggested, including chemical sputtering involving gas-phase ions generated by electrospray ionization (ESI) and subsequent charge transfer between these primary ions and sample molecules on the surface. Also, multiply charged solvent droplets probably impact the surface and dissolve sample molecules, resulting in the formation of charged droplets carrying the sample molecules (ESI-like droplet pick-up mechanism), which then leads to formation of gas-phase sample ions through mechanisms similar to normal ESI.

DESI enables rapid, high-throughput detection of analytes present under ambient conditions on a variety of surfaces positioned between the solvent spray and the MS source. DESI is applicable to small organic molecules as well as to proteins and other biological macromolecules. DESI applications described for drug analysis include direct detection of various drugs in tablets *(8-12)*, ointments applied to the surface of a special holder or a cardboard piece *(8-10)*, liquid pharmaceutical preparations *(10)*, gel formulations applied to the surface of human skin *(10)*, biological fluids (urine, blood, or plasma) absorbed onto a filter paper or dried on other appropriate surface *(7, 9, 10)*, or pharmaceutical samples (components of an over-the-counter pain medication) separated on thin-layer chromatography plates *(13)*.

Desorption atmospheric pressure chemical ionization (DAPCI) *(14)* was developed as an alternative approach for compounds with a lower ionization efficiency in DESI, such as for weakly polar corticosteroids *(9, 10)*. DAPCI employs gaseous ions of volatile compounds (such as toluene or methanol) produced by corona-discharge ionization of vapors carried by a high-velocity nitrogen jet *(14)*. A similar technique, called atmospheric-pressure solids analysis probe (ASAP), was recently reported for a rapid analysis of volatile and semi-volatile liquid or solid materials *(15)*. In ASAP, the sample is vaporized in the hot nitrogen gas stream followed by ionization of the vapors by corona discharge.

Direct Analysis in Real Time (DART)

DART is a new ion source for rapid, non-contact analysis at atmospheric pressure. In DART, the sample surface is exposed to a stream of excited gas, such as metastable helium atoms or nitrogen molecules, which are produced in a discharge chamber of the DART source *(6)*. Several ionization mechanisms have been postulated, including Penning ionization by energy transfer from an excited atom or molecule of energy greater than the ionization energy of the sample molecules. When helium is used, the excited helium neutrals primarily react with atmospheric water to produce hydronium ions, which transfer protons to the sample molecules. As opposed to DESI, DART does not use any solvent, but its applicability is probably limited to smaller organic molecules.

DART has been tested in a wide variety of applications *(6, 10, 16)*, including drug analysis in tablets, capsules, ointments, liquid pharmaceutical preparations, biological fluids (urine, blood, and saliva) or, in the case of illicit drugs, on currency, clothes, glassware or in alcoholic beverages. In the case of fluids, the sample was absorbed onto a filter paper *(10)* or a glass rod was simply dipped into the sample *(6)* and placed in front of DART ionizing beam (the latter procedure has already been automated).

Imaging Mass Spectrometry

Imaging MS produces molecular images of samples through ionization from a clearly identified point on a flat sample and performing a raster of the sample by moving the point of ionization over the sample surface. The collected data (positional data and *m/z* intensities) are converted into images that show distribution of targeted compounds (based on their specific *m/z*) in tissues or in various other materials.

Imaging MS requires direct sampling of the compounds from precisely defined points on the sample surface with an adequate spatial resolution. Spatial resolution refers to the smallest distance between two points that can be clearly distinguished. The most important parameter determining spatial resolution is the size of the focused spot from which the ions are emitted *(17)*.

Two desorption/ionization methods have been mostly employed for MS imaging: (i) SIMS for small (<1000 Da) molecules and (ii) MALDI for larger (<100 000 Da) molecules *(17)*. DESI also has already been tested for this kind of application *(7, 18)*, achieving a spot size of 50 μm. In SIMS, the sample surface is bombarded by a high-energy ion (*e.g.* Ar^+, Cs^+, or Ga^+) beam, which can be focused to spot sizes as small as 10 nm in diameter, although spot sizes of 0.1-30 μm are usually used *(17, 19)*. In MALDI, laser beams can be focused to about 1 μm spots, but much larger spot sizes (5-100 μm in diameter) are more common in practice *(17)*.

A time-of-flight (TOF) mass analyzer is typically used in MS imaging applications both in combination with SIMS or MALDI. TOF is a non-scanning instrument, which provides full spectrum data at fast acquisition rates and often high-resolution and high-accuracy MS measurements. A tandem MS instrument, such as an ion trap or triple quadrupole, also offers enhanced selectivity (as compared to a single-stage MS instruments) and may serve as a lower-cost alternative to a high-resolution TOF.

TOF-SIMS imaging has been employed for the mapping of sample surfaces in a variety of applications *(19)*, including the characterization of drug delivery systems *(20)*. The entire cross-section of a pellet, bead, or capsule containing an active drug encapsulated in a multi-layer coating can be studied to provide information about the drug distribution and the chemical composition and morphology of the coating layers *(21)*. Similarly, TOF-SIMS can be used for imaging of active molecules stamped onto polymer surfaces *(22, 23)*. This kind of characterization can support development of effective drug delivery systems, especially in the case of controlled-release systems, as well as serve for defending against patent infringement and counterfeiting.

MALDI-TOF is particularly useful for imaging of larger molecules, such as peptides and proteins, but smaller molecules can be also analyzed by this technique. Drug researchers are mainly interested in distribution of investigated drugs in target tissues and in tissue response to the presence of the drug, which can be characterized by the production of a specific peptide or protein.

Figure 2 illustrates the basic process involved in an imaging MALDI-TOF MS analysis of biological tissues *(24)*. The sample surface (in this case a rat brain section) is coated with a suitable matrix compound (*e.g.* sinapic acid) that assists in ionization of sample molecules desorbed by a laser. The figure shows an MS spectrum obtained from one, precisely located spot on the sample surface. The peaks in the spectrum represent various compounds and their relative concentrations. Combining these MS data from all the spots, molecular images of different compounds (characterized by a specific *m/z*) can be obtained as demonstrated for the three selected peaks.

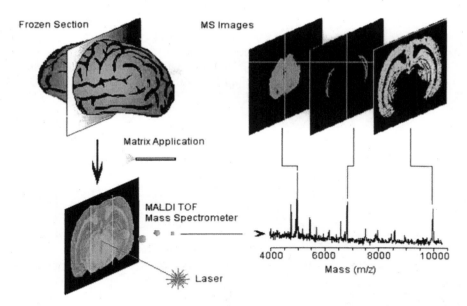

Figure 2. Schematics of the imaging MS process using MALDI-TOF MS (Reproduced with permission from reference 24. Copyright 2001 Nature Publishing Group.)

Mass Spectrometry Combined with High-Field Asymmetric Waveform Ion Mobility Spectrometry (FAIMS)

FAIMS is a relatively new technology for separation of gas-phase ions at atmospheric pressure *(25, 26)*. In FAIMS, a mixture of ions is introduced between two metal plates, to which an appropriate voltage is applied, causing some ion types to drift and hit the metal plates while other types of ions remain between the plates and can reach MS for further separation and detection. The ion drift towards a plate is caused by the difference in ion mobility in strong and weak electric fields, which are applied using a high frequency asymmetric waveform characterized by a significant difference in voltage in the positive and negative polarities of the waveform.

For example, this waveform can be a square wave, in which a high positive voltage is applied for a short time and a low negative voltage is applied for a longer time as shown in Figure 3. The strong field is provided by the application of the peak voltage of the waveform called the dispersion voltage (DV). The weak field of opposite polarity is applied for a correspondingly longer time.

290

When the ion is driven by a strong electric field, the collision of the ion with a bath gas (*e.g.* nitrogen) is more energetic than in the case of a stationary ion. This may increase or decrease ion mobility relative to the mobility of a weak field. The change in mobility is both ion and bath gas dependent. Figure 3 illustrates the motion of an ion with a higher mobility in the strong field.

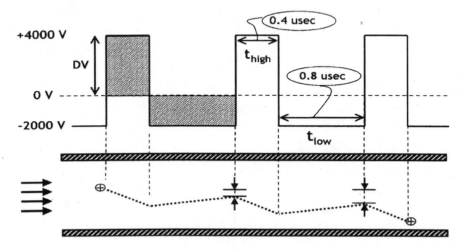

Figure 3. Illustration of the ion motion between FAIMS electrodes for an ion with increasing mobility in strong electric fields. (Reproduced with permission from reference 26. Copyright 2004 Elsevier.)

To stop the ion drift, a small dc voltage of appropriate magnitude and polarity can be applied to either of the plates. This voltage, called the compensation voltage (CV), enables transmission of ions through the FAIMS device, for which the CV is the right voltage to balance the drift caused by the application of the asymmetric waveform of a given DV.

As a result, FAIMS can greatly enhance selectivity by separating analyte ions from chemical background noise using CVs characteristic for given analytes. Also, cylindrical geometry FAIMS can focus and trap ions to provide an increase in sensitivity. These two factors can greatly improve analyte detection in complex matrices by direct MS analysis (or in combination with LC-MS), which was for example demonstrated in the analysis of various drugs in urine *(27)*.

Liquid Chromatography-Mass Spectrometry

In recent years, LC-MS has become a widely used, reliable technique that has been utilized in numerous applications (28). Most modern, commercially available LC-MS instruments employ atmospheric pressure ionization (API), mainly ESI or atmospheric pressure chemical ionization (APCI).

ESI enables analysis of polar compounds, while APCI serves as an alternative for less polar, more volatile molecules that are not as sensitive with ESI. As opposed to APCI, ESI can produce multiply charged ions, thus reduce m/z ratio for large molecules and enable their analysis using conventional MS instruments. Recently introduced atmospheric pressure photoionization (APPI) provides an option for even less polar, weakly or non-ionized compounds, such as steroids (29).

API techniques are soft ionization methods, producing predominantly protonated or deprotonated molecular ions (without or with adducts), which provide limited information about the analyte structure. This disadvantage is in practice usually overcome by using MS^n (mostly MS/MS) or high-resolution, accurate-mass MS that improve detection selectivity and enable further characterization for identification/confirmation purposes. A wide range of tandem MS instruments became available for LC-MS in the past years, including triple quadrupoles (usually the best choice for quantitative analysis), traditional and linear ion traps, or combinations of a quadrupole with a TOF MS. Also, several vendors offer high-resolution, accurate-mass benchtop TOF MS instruments.

The capability of tandem MS or accurate mass measurements to identify a truly unknown compound are, however, far inferior to those of a GC-MS with electron ionization (EI). EI spectra have characteristic fragmentation patterns, which can be interpreted to elucidate compound structures or searched against a database (MS library). Although some MS/MS or accurate-mass libraries exist for LC-API-MS, they can hardly be compared to extensive, (practically) instrument-independent EI spectra libraries.

Particle beam LC-MS can provide EI spectra but this technique became almost forgotten because of its rather low sensitivity and limited applicability compared to LC-MS with ESI or APCI. Recently reported LC-MS with EI of cold molecules in supersonic molecular beams (SMB) has a potential to provide an invaluable identification tool for a wide range of small organic molecules (30), but this technique has not been commercialized yet. SMB-MS has been successfully combined with GC to offer many unique features as discussed in the section on GC-MS.

The most prominent trend, observed basically in any analytical field, is the effort to speed the analysis and provide high throughput measurements, which are very useful in pharmaceutical applications (*e.g.* in drug discovery, development, or quality control). In fast LC-MS, the LC part dictates the speed, while the MS must keep up with it, which mainly means that the MS instrument should provide fast data acquisition rates to detect rapidly eluting peaks from the LC column.

In addition to employing shorter and narrower (packed or capillary) columns, two main fast LC approaches are currently investigated and start being applied in practice: the use of columns with small particle sizes (< 2 μm) and column operation at elevated temperatures.

Fast Liquid Chromatography using Small Particle Sizes

According to the chromatographic theory, as the particle size decreases, the separation efficiency increases because the number of theoretical plates is inversely proportional to the particle size, and so is the optimum linear velocity of the mobile phase. Thus, columns with smaller particle sizes can provide an increase in speed due to higher optimum flow rates. Although this theory has been well known for decades, the practical application of this approach has been difficult for several reasons.

For one, the operation of small particle size columns at high flow rates creates high back pressures (often exceeding 10,000 psi), requiring special instrumentation that enables injection of samples and reproducible mobile phase pumping at these pressures. Moreover, the entire LC system must have very low extra-column volumes to preserve the separation efficiency. Another challenge involves the design and development of < 2 μm particles and their packing into reproducible and rugged columns. Also, the samples must be well-filtered to prevent clogging of the frits holding the particles in the column.

Recently, two manufacturers introduced columns with smaller particles (1.7 and 1.8 μm) together with compatible LC-MS systems, providing adequate pressure limits, injection systems, low extra-column volumes, as well as fast MS detection for narrow peaks. One manufacturer termed the technology as ultra performance liquid chromatography (UPLC) *(31)*, whereas the other describes the columns as suitable for rapid resolution LC.

The UPLC term and technology is spreading throughout the scientific community, which can be demonstrated by a number of papers evaluating this approach in various applications, including metabolite profiling in urine or serum *(32-35)*, drug development studies *(36)*, forensic drug analysis *(37)* or multiresidue analysis of veterinary drugs in milk and eggs *(38)*.

High-Temperature Liquid Chromatography

The application of elevated temperatures in high-temperature LC (HTLC) decreases viscosity of the mobile phase and increases analyte diffusivity, which leads to lower back pressures and increased mass transfer, respectively *(39)*. As a result, LC columns can be operated at higher flow rates, reducing the analysis time without making significant sacrifices in the separation efficiency. Moreover, elevated temperatures can be employed in the combination with small particle size columns, thus providing an additional gain in speed or separation efficiency, while keeping the backpressure within reasonable limits. Also, temperature programming of the column can change separation selectivity, which may improve analyte separation or serve as an alternative to solvent gradient.

Another attractive aspect involves the use of superheated water (at 100-200°C) as a potential replacement for medium-polarity mobile phases (such as acetonitrile-water mixtures) in reversed-phase LC *(40, 41)*. This would reduce solvent and waste costs and simplify system operation. Under high-temperature conditions, water behaves as a moderate polarity solvent because its dielectric constant decreases from about 80 to about 35 over the range of 25 to 200°C *(42)*, while it retains an appreciable density (> 0.85 g/mL), cohesive energy, and hydrogen bonding potential *(43)*.

Until recently, the application of HTLC was rather difficult in practice due to instrument and column limitations. Instrumentation for HTLC is now available, which allows operation at temperatures up to 200°C with mobile phase preheating to eliminate thermal mismatch *(44)*. The thermal mismatch occurs when the cool mobile phase enters the heated column and warms up faster along the walls, leading to band broadening due faster flow of the mobile phase along the column wall than in the center *(45)*.

Another practical concern involves the stability of the stationary phase at the elevated temperatures. Conventional, reversed-phase LC columns with silica-based stationary phases are stable at temperatures up to 90°C (50-60°C limit is more common). Stationary phases with higher temperature stability are based mainly on zirconium oxide, although graphitized carbon or rigid polystyrene-divinylbenzene phases can also tolerated elevated temperatures *(44)*.

With the recent advancements in column technology and instrumentation, HTLC may soon become a practical tool, especially in the combination with narrower columns and/or columns with small particle sizes and in applications that can tolerate higher temperature with respect to the analyte stability.

Gas Chromatography-Mass Spectrometry

GC-MS is a suitable technique for thermally stable compounds that can be readily volatilized. Most drugs are rather polar molecules, thus require derivatization prior to the GC-MS analysis. The availability and affordability of LC-MS instruments have led to the replacement of GC-MS in most drug analyses in modern laboratories. However, GC-MS with derivatization may still be employed for confirmation purposes, providing orthogonal selectivity to LC-MS. Also, certain coumpounds, such as steroids, do not provide sufficient sensitivity in common LC-API-MS techniques, thus GC-MS with derivatization remains a viable approach in these cases (46).

The derivatization procedure can be automated using, e.g. in-vial derivatization, on-fibre derivatization (47, 48) or derivatization performed directly in the GC inlet. The latter approach can be done using the direct sample introduction (DSI) technique (49). In DSI, a liquid or solid sample is placed in a disposable microvial, to which a derivatization solution can be added (50). After this step, the microvial is introduced into the injection port using a manual probe or more recently using an autosampler. In the automated version, the liquid sample and/or derivatization solutions are injected into the microvial placed in a liner (51), which is then inserted into the inlet (or a thermodesorption unit attached to the inlet).

Some fast GC-MS techniques may permit analysis of less volatile and thermally stabile analytes even without derivatization (52). Supersonic molecular beam GC-MS (GC-SMB-MS) is of a particular interest in this respect. In GC-SMB-MS, organic molecules pass through a small opening (about 0.1 mm i.d.) placed between the GC outlet and the MS, form an SMB by co-expansion with a carrier gas into vacuum and are vibrationally supercooled in the process (53). As a result, an enhanced molecular ion occurs practically for all molecules, increasing the MS detection selectivity in EI. SMB-MS still provides typical EI fragmentation patterns, thus enabling conventional MS library searching, but with a higher confidence in correct compound identification due to the presence of a prominent molecular ion.

SMB-MS can handle very high flow rates (up to 240 ml/min with the prototype instrument), enabling very fast analyses, especially when combined with a short megabore column. In addition to an increase in speed, the fast flow rates and other unique features of GC-SMB-MS also enable analysis of thermally labile and low-volatility analytes, such as large polycyclic aromatic hydrocarbons or drug molecules (see Figure 4), thereby extending the scope of GC-MS to compounds currently done only by liquid chromatography (54, 55). However, the lack of commercial availability is currently a severe limitation in the use of the GC-SMB-MS approach.

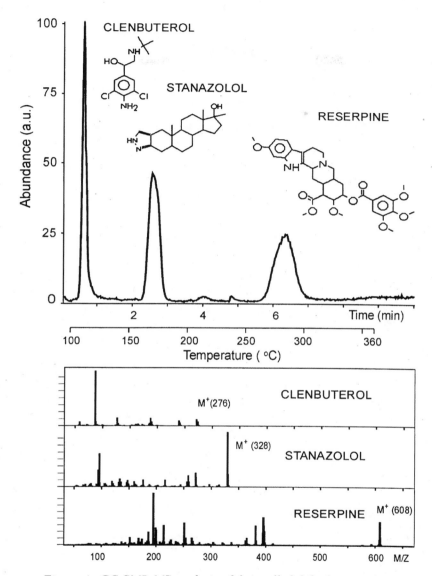

Figure 4. GC-SMB-MS analysis of thermally labile drugs without derivatization. (Reproduced with permission from reference 55. Copyright 2003 Elsevier.)

References

1. Karas, M.; Hillenkamp, F. *Anal. Chem.* **1988**, *60*, 2299-2301.
2. Fenn, J. B. Mann, M.; Meng, C. K.; Wong, S. F.; Whitehouse, C. M. *Science* **1989**, *246*, 64-71.
3. Laiko, V. V.; Baldwin, M. A.; Burlingame, A. L. *Anal. Chem.* **2000**, *72*, 652-657.
4. Laiko, V. V.; Moyer, S. C.; Cotter, R. J. *Anal. Chem.* **2000**, *72*, 5239-5243.
5. Takats, Z.; Wiseman, J. M.; Gologan, B.; Cooks, R. G. *Science* **2004**, *306*, 471-473.
6. Cody, R. B.; Laramee, J. A.; Durst, H. D. Anal. Chem. **2005**, *77*, 2297-2302.
7. Takats, Z.; Wiseman, J.M.; Cooks, R.G. *J. Mass Spectrom.* **2005**, *40*, 1261-1275.
8. Chen, H.; Talaty, N.N.; Takats, Z.; Cooks, G.R. *Anal. Chem.* **2005**, *77*, 6915-6927.
9. Williams, J.P.; Scrivens, J.H. *Rapid Commun. Mass Spectrom.* **2005**, *19*, 3643-3650.
10. Williams, J.P.; Patel V.J.; Holland, R.; Scrivens, J.H. *Rapid Commun. Mass Spectrom.* **2006**, *20*, 1447-1456.
11. Rodriguez-Cruz, S. E. *Rapid Commun. Mass Spectrom.* **2006**, *20*, 53-60.
12. Leuthold, L. A.; Mandscheff, J. F.; Fathi, M.; Giroud, C.; Augsburger, M.; Varesio, E.; Hopfgartner, G. *Rapid Commun. Mass Spectrom.* **2006**, *20*, 103-110.
13. Van Berkel, G. J.; Ford, M. J.; Deibel, M. A. *Anal. Chem.* **2005**, *77*, 1207-1215.
14. Takats, Z.; Cotte-Rodriguez, I.; Talaty, N. N.; Chen, H. W.; Cooks, R. G. *Chem. Commun.* **2005**, *15*, 1950-1952.
15. McEwen, C. N.; McKay, R. G.; Larsen, B. S. *Anal. Chem.* **2005**, *77*, 7826-7831.
16. http://www.jeol.com/ms_/msprods/accutof_dart.html
17. Todd P. J.; Schaaff, T. G.; Chaurand, P.; Caprioli, R. M. *J. Mass Spectrom.* **2001**, *36*, 355-369.
18. Wiseman, J. M.; Puolitaival S. M.; Takats, Z.; Cooks, R. G.; Caprioli, R. M. *Angew. Chem. Int. Ed.* **2005**, *44*, 7094-7097.
19. Belu, A. M.; Graham, D. J.; Castner, D. G. *Biomaterials* **2003**, *24*, 3635-3653.
20. Davies, M. C.; Shakesheff, K. M.; Roberts, C. J.; Tendler, S. J. B.; Bryan, S. R.; Patel, N. In *The Encyclopedia of Drug Delivery*; Mathiowitz, E., Ed.; Wiley: New York, NY, 1999; 269-277.
21. Belu, A. M.; Davies, M. C.; Newton J. M.; Patel, N. *Anal. Chem.* **2000**, *72*, 5625-5638.

22. Yang, Z.; Belu, A. M.; Liebmann-Vinson, A.; Sugg, H.; Chilkoti, A. M. *Langmuir* **2000**, *16*, 7482-7492.

23. Hyun, J.; Zhu, Y.; Liebmann-Vinson, A.; Thomas, P.; Beebe, J. *Langmuir* **2001**, *17*, 6358-6367.

24. Stoeckli, M.; Chaurand, P.; Hallahan, D. E.; Caprioli, R. M. *Nature Medicine* **2001**, *7*, 493-496.

25. Guevremont, R.; Purves, R. W. *Rev. Sci. Instrum.* **1999**, *70*, 1370-1383.

26. Guevremont, R. *J. Chromatogr. A* **2004**, *1058*, 3-19.

27. McCooeye, M. A.; Mester, Z.; Ells, B.; Barnett, D. A.; Purves, R. W.; Guevremont, R. *Anal. Chem.* **2002**, *74*, 3071-3075.

28. Niessen, W. M. A. *J. Chromatogr. A* **2003**, *1000*, 413-436.

29. Robb, D. B.; Covey, T. R.; Bruins, A. P. *Anal. Chem.* **2000**, *72*, 3653-3659.

30. Granot, O.; Amirav, A. *Int. J. Mass Spectrom.* **2005**, *244*, 15-28.

31. Swartz, M. E. *J. Liq. Chrom.* **2005**, *28*, 1253-1263.

32. Wilson, I. D.; Nicholson, J. K.; Castro-Perez, J.; Granger, J. H.; Johnson, K. A.; Smith, B. W.; Plumb, R. S. *J. Proteome Res.* **2005**, *4*, 591-598.

33. Crockford, D. J.; Holmes, E.; Lindon, J. C.; Plumb, R. S.; Zirah, S.; Bruce, S. J.; Rainville, P.; Stumpf, C. L.; Nicholson, J. K. *Anal Chem.* **2006**, *78*, 363-371.

34. O'Connor, D.; Mortishire-Smith, R.; Morrison, D.; Davies, A.; Dominguez, M. *Rapid Commun. Mass Spectrom.* **2006**, *20*, 851-857.

35. Nordstrom, A.; O'Maille, G.; Qin, C.; Siuzdak, G. *Anal Chem.* **2006**, *78*, 3289-3295.

36. Wren, S. A.; Tchelitcheff, P. *J. Chromatogr A.* **2006**, *1119*, 140-146.

37. Apollonio, L. G.; Pianca, D. J.; Whittall, I. R.; Maher, W. A.; Kyd, J. M. *J Chromatogr. B* **2006**, *836*, 111-115.

38. Cui, X.; Shao, B.; Zhao, R.; Yang, Y.; Hu, J.; Tu, X. *Rapid Commun. Mass Spectrom.* **2006**, *20*, 2355-2364.

39. Greibrokk, T.; Andersen, T. *J. Chromatogr. A* **2003**, *1000*, 743-755.

40. Fields, S. M.; Ye, C. Q.; Zhang, D. D.; Branch, B. R.; Zhang, X. J.; Okafo, N. *J. Chromatogr. A* **2001**, *913*, 197-204.

41. Coym, J. W.; Dorsey, J. G. *J. Chromatogr. A* **2004**, *1035*, 23-29.

42. Yang, Y.; Belghazi, M.; Lagadec, A.; Miller, D. J.; Hawthorne, S. B. *J. Chromatogr. A* **1998**, *810*, 149-159.

43. Pawlowski, T. M.; Poole, C. F. *Anal. Commun.* **1999**, *36*, 71-75.

44. Marin, S. J.; Jones, B. A.; Felix, W. D.; Clark, J. *J. Chromatogr. A* **2004**, *1030*, 255-262.

45. Mao, M.; Carr, P. *Anal. Chem.* **2001**, *73*, 4478-4485.

46. Van Thuyne, W.; Delbeke, F.T. *Biomed Chromatogr.* **2004**, *18*, 155-159.

47. Rodríguez, I.; Carpinteiro, J.; Quintana, J. B.; Carro, A. M.; Lorenzo, R. A; Cela, R. *J. Chromatogr. A* **2004**, *1030*, 255-262.

48. Chia, K.-J.; Huang, S.-D. *Anal. Chim. Acta* **2005**, *539*, 49-54.

49. Amirav, A.; Dagan, S. *Eur. Mass Spectrom.* **1997**, *3*, 105-111.
50. Ding, W. H.; Liu, C. H.; Yeh, S. P. *J. Chromatogr. A* **2000**, *896*, 111-116.
51. Cajka, T.; Mastovska, K; Lehotay, S. J.; Hajslova, J. *J. Sep. Sci.* **2005**, *28*, 1048-1060.
52. Mastovska, K; Lehotay, S. J. *J. Chromatogr. A* **2003**, *1000*, 153-180.
53. Amirav, A.; Gordin, A.; Tzanani, N. *Rapid Commun. Mass Spectrom.* **2001**, *15*, 811-820.
54. Dagan, S.; Amirav, A. *J. Am. Soc. Mass Spectrom.* **1996**, *7*, 737-752
55. Fialkov, A.B.; Gordin, A.; Amirav, A. *J. Chromatogr. A* **2003**, *991*, 217-240.

Indexes

Author Index

Subject Index